Cover design by Tulasi Zimmer

Cover photograph by Patrick N. Smith, PLS

ISBN: 978-0-9888737-1-1

Publication date January 10 2013

Published by Montana Technical Writing

Preface

The United States Department of the Interior Bureau of Land Management created the Geographic Coordinate Database in order to provide agencies with a unified spatial framework for aligning cadastral GIS information across the Western US in a consistent way. In April 2000 the Western Governors Association adopted the Bureau of Land Management's Geographic Coordinate Database (GCDB) as the preferred representation of the Public Lands Survey System (PLSS) for GIS applications. This is significant in the western states where most land ownership, stewardship boundaries, and administrative boundaries are based on or referenced to the PLSS. That means that any GIS data of those boundaries must use the PLSS as the mapping framework.

It is important to standardize on a single representation of the PLSS in order to integrate GIS and mapping data that come from a variety of sources. The GCDB, as implemented by the BLM and as adopted by the Western Governors Association does meet that requirement, and the GCDB is now available for most of the western states.

The purpose of this GCDB handbook is to provide a quick reference to the practical applications of the Bureau of Land Management's Geographic Coordinate Database for Geographic Information System users and land surveyors. Because the GCDB is important to GIS in the Public Lands Survey states, GIS users and land surveyors should understand what the GCDB is and how the GCDB affects GIS data alignment and spatial accuracy, and should know how to use the GCDB for a number of GIS and surveying tasks.

The author hopes that understanding will lead to correct use and application of the data.

How to use this handbook.

This handbook may be read non-sequentially. The reader may directly jump to most any section without having to read prior chapters. Each chapter is roughly self-contained. To achieve that independence, some content may repeat in order to provide relevant context, although major topics may refer to other sections of the book.

Additional training resources, including video tutorials, are available on the website GIS for Surveyors at http://gisforsurveyors.com/gcdb_training.htm.

Please send comments and/or corrections to RjZimmer@GISforSurveyors.com

Table of Contents

Chapter 1

Ch. 1 About the PLSS and the GCDB

In this chapter, we take a brief look at the history of the Public Land Survey System, how it relates to Geographic Information Systems, and the rationale for creating the Geographic Coordinate Database.

Overview of the Public Lands Survey System of the Western United States.

Thomas Jefferson devised the Public Lands Survey System (PLSS) which the US congress implemented in 1785 as a means to survey and to describe lands for sale by the U.S. federal government. Most of the lands in the western United States are PLSS states, so there is a heavy reliance on the PLSS for surveying and mapping.

From the Bureau of Land Managements' 1947 Manual of Surveying Instructions: ...(a) ...*the public lands of the United States shall be divided by lines intersecting true north and sought at right angles so as to form townships [that are]6 miles square ,(b) the townships shall be marked with progressive numbers from the beginning, (c) the townships shall be subdivided into 36 sections, each 1 mile square and containing 640 acres as nearly as may be, and (d)the sections shall be numbered, respectively, beginning with number 1 in the northeast section, and proceeding west and east alternately through the township with progressive numbers to and including 36.*

PLSS sections may be further subdivided subsequent to the initial survey by using a system of aliquot parts of geometric division by midpoints. Thus a section may be subdivided into four quarters, each of approximately 160 acres, and each quarter-section may be further subdivided into quarter-quarter sections of approximately 80 acres, and on down. The aliquot parts are described by location

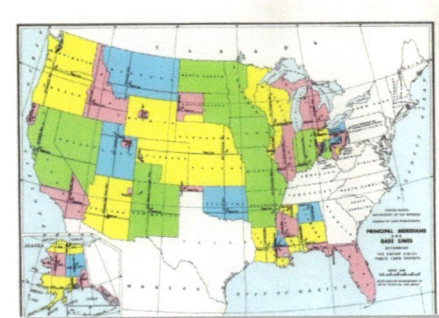

Figure 1 Map of the PLSS States - from the US Dept. of the Interior 1988

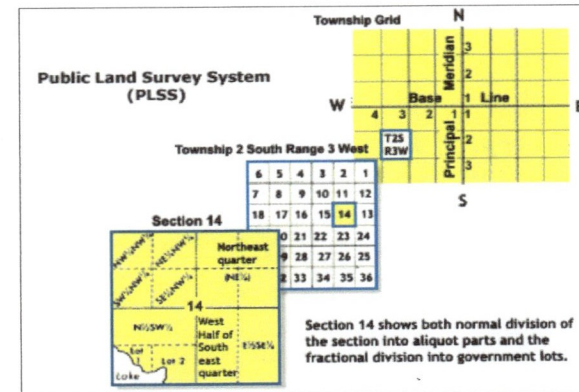

Figure 2 PLSS Subdivisions - from the US Dept. of the Interior 1988

within the section and their degree of division.

The PLSS is the legal framework upon which most real property descriptions are based in the US western states. In addition to real property, other legal descriptions such as administrative areas (fire districts, zoning, taxing districts, and others) may be based on the PLSS. Even when the legal description may be metes and bounds type of description the location of the property will be referenced to the PLSS as shown in the example. The highlighted text (...*in the Northwest Quarter [of] Section 19, Township 10 North, Range 3 West, Principal Meridian [of] Montana*...) indicates where in the US this particular property is located. The description then goes on to describe the geometry of the property in its particulars.

PERIMETER DESCRIPTION—COMBINED

A TRACT OF LAND IN NORTHWEST QUARTER SECTION 19, TOWNSHIP 10 NORTH, RANGE 3 WEST, PRINCIPAL MERIDIAN MONTANA, LEWIS AND CLARK COUNTY, MONTANA, AND BEING COMPRISED OF FOUR GEORGIANS SCHOOL PROPERTY AND LOTS 30A AND 30B OF CLOVERVIEW PUD, AND PUBLIC ROAD RIGHT—OF—WAY, LOCATED IN THE CITY OF HELENA, AND FURTHER DESCRIBED BY METES AND BOUNDS AS FOLLOWS:

BEGINNING AT THE NORTHEAST CORNER OF SAID LOT 30A, CLOVERVIEW P.U.D.; THENCE S00°09'53"W, 295.26 FT. ALONG THE WEST RIGHT—OF—WAY LINE OF McHUGH LANE TO A POINT BEING THE NORTHEAST CORNER OF LOT 31A CLOVERVIEW P.U.D ON THE SOUTH RIGHT—OF—WAY LINE OF A 50 FT. WIDE PUBLIC RIGHT—OF—WAY; THENCE ALONG SAID RIGHT—OF—WAY, S89°22'52"W, 246.57 FT. TO THE END OF SAID RIGHT—OF—WAY; THENCE S18°49'09"W, 114.26 FT. TO THE SOUTHEAST CORNER OF THE FOUR GEORGIANS SCHOOL PROPERTY; THENCE ALONG THE SOUTH LINE OF SAID SCHOOL PROPERTY, N80°12'18"W, 100.00 FT.; THENCE N18°34'25"W, 54.98 FT.; THENCE 204.20 FT. ON A CURVE TO THE LEFT, WITH A RADIUS OF 130.00 FT., AND A CHORD THAT BEARS N63°34'25"W, 183.85 FT.; THENCE S71°25'35"W, 740.26 FT. TO THE SOUTHWEST CORNER OF SAID SCHOOL PROPERTY; THENCE ALONG THE WEST LINE OF SAID SCHOOL PROPERTY, N00°39'42"W, 477.26 FT. TO THE NORTHWEST CORNER OF SAID SCHOOL PROPERTY ON THE SOUTH RIGHT—OF—WAY—LINE OF CUSTER AVENUE; THENCE ALONG SAID RIGHT—OF—WAY, N89°23'00"E, 1272.24 FT. TO THE NORTHEAST CORNER OF SAID LOT 30A THE POINT OF BEGINNING.

Figure 3 Example legal description with a PLSS location

Introduction to the Geographic Coordinate Database – the digital version of the PLSS

In the early decades of GIS data development, people who needed a PLSS framework for digitizing parcels and boundary data devised and developed many representations of the PLSS, because there was no readily available, consistent, unitary dataset for the PLSS. Beginning in the 1990's when various agencies and commercial concerns started to create their own digital PLSS data sets, they typically digitized the US Geological Survey (USGS) topographical map series which show PLSS cartography. Additionally, the US Census Bureau created its version of the PLSS which it included in its Topologically Integrated Geographic Encoding and Referencing system (TIGER) files for the 1990 census. In all cases the early PLSS datasets were manually digitized from hard copy maps using only a few control points. Thus the spatial accuracy of the early PLSS datasets were driven by the accuracy of the hand drafted original maps then further degraded by the vagaries of the digitizing process. Therefore, the spatial accuracy of each PLSS dataset varied considerably within the dataset, and was usually not sufficiently accurate for densely populated areas. Additionally, competing PLSS datasets did not spatially align with each other.

Due to accuracy issues, lack of availability, the cost of acquiring reference data sets, and other reasons, many GIS data conversion project managers created their own project specific PLSS framework for mapping their data. The result is that there are many GIS datasets today based on non-conforming PLSS. Thus, when people bring other datasets together, the GIS geometries do not align where they should, for example neighboring counties' boundaries and parcels data may not align at the edges.

The vision for the GCDB was to standardize on a reliable and consistent digital PLSS which would be as accurate as practicable and which could improve as time and money allowed.

How the GCDB is built

Early on, the BLM recognized the need for a single accurate source of the PLSS that could be used for a variety of purposes by various agencies, as well as the surveying and GIS communities. Starting in 1989 the BLM began creating the GCDB. The Bureau of Land Management designed a system of using the best available survey measurements of the lines between points of the PLSS to compute each point's geographic coordinates. For that purpose the GCDB was created from the original survey notes of the GLO or other record information which describe the bearings and distances between PLSS corners. T

The first step was to obtain highly reliable coordinates on *control points* of the PLSS such as township corners, then calculate the coordinates for other PLSS points based on the historic, or when available, more recent survey measurements between points. This method created points and lines between points .The process begins at a corner with known coordinates (latitude and longitude), and uses the bearings and distances from corner to corner, in order to derive coordinates for each corner. The process also captures other relevant information such as the record source (whether GLO or survey record number, surveyor name, date of survey, etc.) and an estimate of the quality (spatial error) of the derived coordinate. The result is a coordinate for each corner and an error estimate for that coordinate. The accuracy of the GCDB is based on the accuracy of the survey information, the level of control and the method of data entry. The accuracy can vary from a few feet to over 300 feet, with ±40 feet being typical. Problem areas can be improved upon by incorporating more recent surveys or providing GPS control.

In some areas the GCDB is built primarily from the GLO notes supplemented with GPS control. Other areas have incorporated the most recent survey information. The inclusion of recent surveys typically improves the accuracy of the corner coordinates. Regardless of the data sources and methods used to create the initial GCDB, the coordinate data can be improved by incorporating new surveys and GPS control then re-adjusting the GCDB. The BLM has procedures to enter new data and re-adjust the township coordinates. The BLM and the GIS community expect that the GCDB will continuously improve over time as new surveys are performed and incorporated into the GCDB. Any adjustments to those data sets that are based on the GCDB will also have to be adjusted - see chapter 4 *How to Improve the GCDB*.

The BLM uses software to perform the computations. The software, called Geographic Measurement Management (GMM), creates a database of survey measurements, control points, data quality estimates, error analysis, and PLSS geometry which combine to create the GCDB and which can also be used to improve the accuracy of the coordinates of the PLSS points.

The Geographic Coordinate Database is the mathematical representation of the PLSS. The GCDB contains points, lines, and polygons and attributes for townships, sections, lots, mineral surveys of the PLSS along with survey measurement data. The GCDB assigns coordinates to the corner points of the PLSS townships, sections, one-quarter sections, one-quarter-quarter (1/16) sections, and any meanders, lots, or mineral survey corners of a township. From these coordinated points the line and polygon geometries are composed. All these features and attributes (some of which are described later) along with metadata on the methods and quality of the data, comprise the GCDB data sets.

Maintenance and Enhancements

Who is responsible for maintaining, and improving the GCDB? Although the BLM had funding for the creation of the GCDB, responsibility and funding for the maintenance and upgrading of the GCDB are unresolved issues. Most likely, the BLM will retain responsibility for the GCDB. That was the recommendation of the Western Cadastral Data and Policy Forum for the Western Governors Association Geographic Information Council. According to the participants of that forum the "GCDB is the best hope of standardizing the PLSS in western states and its use is strongly endorsed by the WGA sponsored Western Cadastral Data and Policy

Forum and the WGA GIS Advisory Council." That group passed a resolution to "...strongly support increased federal funding for continued collection, accuracy and content enhancement and distribution of GCDB."

During 2010, and 2011 the BLM contracted for GCDB spatial accuracy enhancements in many western states (Oregon, Colorado, Oregon, etc.) in order to support energy development programs. The BLM however, is not the only party involved with the GCDB. Although the BLM currently is responsible for creating the GCDB, the surveying community can actively participate in enhancing the quality of the GCDB by sharing geographic coordinate information on the PLSS, as well as providing upgraded corner ties to section, township, meander, and other PLSS corners to the BLM. The BLM can then incorporate these improved data and perform a readjustment of the GCDB. The result is a GCDB that better serves the surveyors' needs as well as the GIS community's. Surveyors can also help the GIS community by performing coordinate adjustments of GIS data sets that depend on the GCDB.

The GCDB Data Model

The GCDB Data Model consists of a set of files for identifying PLSS corners, storing: raw survey measurements, control point files; adjusted coordinates files; comments; and error reporting files. The GIS data model for the GCDB contains points, lines, and polygons together with various attributes that represent the various components of the digital PLSS – corners, township and section lines, township and section polygons, etc.

The GCDB contains points for corners of townships, sections, and quarter sections, and may contain other points (e.g. 1/16 corners, or meander corners). The GCDB *points* serve as the control for registering GIS layers to the PLSS. That is, geometries can be snapped (made geometrically coincident) to the GCDB. Alternatively, the GCDB line and polygon geometry can be used as GIS primitives from which to construct more complex geometries. GCDB points are also the basis for doing spatial accuracy improvement adjustments – i.e. the points form the control reference points. Each GCDB point has a unique identifier attribute (PLSSID) which is universally unique for every point in the nation.

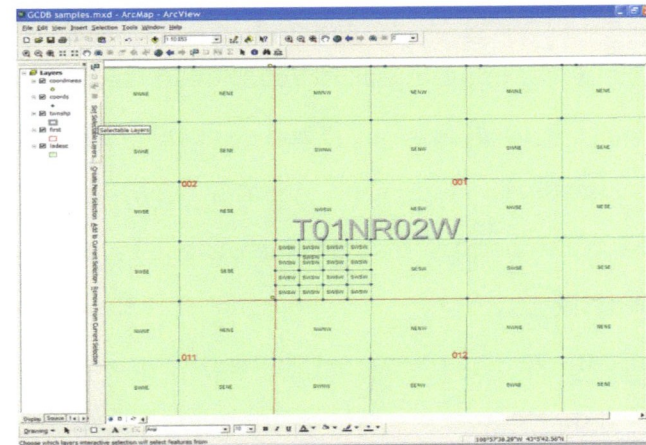

Figure 4 Example GCDB cartography showing geometry and labels

GCDB Topology

The GCDB GIS database design supports topology with the GCDB datasets. Additionally, the GCDB supports topological relationships among GIS data sets. Topology rules can help ensure data integrity and can be used to highlight errors in GIS geometries.

Overview

Topology is the branch of mathematics concerned with those properties of objects that do not change when the object's geometry is twisted, stretched or deformed. According to Merriam-Webster, topology is the branch of mathematics concerned with those properties of geometric configurations (as point sets) which are unaltered by elastic deformations (as a stretching or a twisting). "Topology can be used to abstract the inherent connectivity of objects while ignoring their detailed form." (Eric W. Weisstein, Encyclopedist, Wolfram Research, Inc.). The relationship between different segments of a stream can be described as a linear network. The topological rules of the stream describe the behavior of the segments with respect to each other, irrespective of the shape of the stream. The rules may include conditions such as - *each segment must connect to one or more segments in sequence, and the flow must go only in one direction*. The topologic rules would apply regardless of any geometric alteration that the stream may undergo. That is, the segments must connect and the flow must go in only one direction whether the geometry of the stream is a straight line or a curved, sinuous line. Changing the shape of the stream does not change the topologic relationships. For GIS purposes, connectivity can exist between line segments, as in a pipeline or stream, or between polygons representing parcels or subdivisions.

Topology is germane to surveying where the concepts are applied frequently, although, perhaps not thought of in those terms. Topologic rules that apply to real property, for example, include conditions such as *parcels must not have gaps or overlaps*, and *a parcel's boundary must form a closed polygon*. These topologic rules apply to parcels regardless of the shape of the parcel, and apply to a parcel even when its shape changes. These are important concepts, which surveyors apply to boundary work. When a boundary line adjustment is made between lots, the topological relationships are maintained: the lots remain contiguous, they do not overlap, and no gaps exist between them.

Within GIS, topology is used to model data, or to model behaviors of GIS objects. GIS applies topology concepts to geometric objects in order to model the behavior of the collections of features in a data set, and to describe the relationships between different data sets. Different types of GIS objects (polygons, lines, etc.) may have different topologic rules, as exhibited above. Additionally, GIS can even model topological relationships between different collections of objects. A subdivision polygon would have a topological relationship to the boundaries of the lots that comprise it, or voting precincts may have topological relationship with census block boundaries.

Topology rules can be enforced during the creation and editing of data sets to ensure the integrity of the relationships. Thus, when parcel lines are edited, for instance (such as for a boundary line adjustment), the topology rules help to avoid errors such as moving one parcel line without moving the adjacent parcel line (avoiding gaps or overlaps). In addition, topology allows behavior modeling for geometric networks, polygon geometries, and other types of GIS objects. The advantages to using topology come when analysis and queries are performed on the data. For example, if topology rules for linear networks are applied to a GIS representation of a sewer system, then the system's flows can be analyzed.

How will data be used?

When developing a GIS or converting data into digital form for GIS, it is always important to understand the questions that the data will need to answer in order to design the data model including the topology rules, to support those questions. Just as there is no one way to build a truck, there is no one way to build a GIS data set. The purpose for which the data are used must be fully understood at the outset, and then appropriate topology rules can be applied. While it may be true that a parcel's polygons must not have gaps or overlaps, polygons of different noxious weeds might overlap and have gaps. When modeling linear networks, different topology rules might apply to water systems than to road networks.

Purpose is important when it comes to parcels representing property boundaries. Surveyors typically are called on to define the location of parcel edges and angle points, that is, the polygon shape is of primary importance. Because surveyors tend to concentrate on the boundary aspects of real property, there is a tendency among surveyors to cringe at anything that does not honor those explicit boundary geometries. Nevertheless, there are far more applications in GIS for which the boundary information is of secondary importance or even completely irrelevant. Therefore, for many GIS applications, changing the boundary geometry does not affect the answers being asked of the data. The mathematics of the boundary, which is the bread and butter of the surveyor, is incidental to most GIS applications, because for most GIS applications, the topologic rules are more important than the boundary geometry. That is why in the GIS world a stretching or twisting of the geometry has little or no effect on the ability to perform the important GIS functions. If, for instance, one is only concerned with the value of land within a certain geographic area, or in creating a list of landowners who will be affected by a road closure, then the shapes of those parcels are only a secondary concern.

When it comes to topology, the main difference between GIS and surveying is a matter of perspective, which stems from purpose. Surveyors often focus on physically and mathematically describing the edges of parcels, that is, the boundaries of objects. Whereas most users of parcel data in GIS are concerned with what applies within that configuration (what it touches, what the parcel is in, what the parcel is composed of, what it contributes to, what is near it, what crosses it, etc.) as well as some information extrinsic to the geometry (who owns it, how much it is worth, etc.). In instances where the surveyor is concerned with who owns a parcel, or what zone it is in, the parcel's edges are not necessarily important either. Yet, when the surveyor must monument the corners of the parcel, or plat a new parcel, then the edges are of primary importance.

The greatest value of the GCDB is as a basis for other GIS layers, such as land ownership boundaries, government unit boundaries, fire, conservation, or school districts or others. Again, the reasons for using the GCDB are to standardize on a common PLSS and to improve the spatial accuracy of the PLSS representation. New GIS data that is based on the GCDB framework should better align with other GIS layers from other sources. In addition, existing GIS data can be adjusted to fit the GCDB.

Topology can help one to manage the relationships between GIS layers, is generated by storing GIS layers in a geodatabase, then establishing the relationships between layers. The GCDB data design supports topology. In the PLSS states, legal descriptions for properties, public or private, as well as government and non-government agency boundaries are based on the PLSS. So in addition to federal lands (e.g. US Forest Service), fire district, school districts, county boundaries, a variety of administrative boundaries, such as census tracts and voting precincts, have some elements of the Public Lands Survey System in their legal descriptions. Thus, these legal descriptions have a topological relationship with the GCDB. That is, the lines of legal descriptions should coincide with the lines of the townships and the lines of the sections or parts of sections. One can build a GIS layer from such legal descriptions by snapping the lines to the coordinated points of the GCDB.

Generating the topological relationships between the GCDB and *existing* GIS layers, is a bit more problematic and can be a tedious undertaking. Making an existing GIS boundary, such as a fire district boundary, conform to the GCDB, may involve editing the vertices of the existing boundary to snap them to the GCDB. Alternatively, some of the snapping could be automated through the construction of a topological relationship between the layers of the GCDB and the boundary layer. However, the results from an automated topological process, where a search radius is generalized for the entire dataset, can yield unpredictable results, so this must be done with care.

For more information on the GCDB file contents, see http://www.blm.gov/wo/st/en/prog/more/gcdb.html . For more information about the GCDB identifier naming schema; GCDB data standards, see the GCDB Resources chapter in this book.

The website GIS for Surveyors has an interactive online tool for identifying the GCDB ID for typical corners in a normal township: http://gisforsurveyors.com/gcdb_training

> Adopting the GCDB as your PLSS standard is a step toward data integration and enhanced data sharing.

Examining the Spatial Accuracy of the GCDB

The GCDB was constructed from a variety of records, including original General Land Office field notes and plats and modern private land surveys. The map controls for the GCDB varied greatly in accuracy, ranging from USGS topographic map coordinates to Global Positioning System control points. Because the accuracy of the source documents varies and the number and accuracy of the controls vary, the accuracy of the coordinates within the GCDB varies greatly as well. Accuracies may range from less than one foot to greater than four hundred feet and are dependent on both the source documents and the availability and quality of control. Therefore, the BLM included coordinate reliability information for each point in the GCDB. The coordinate reliability (or simply *reliability*) informs the user of the estimated accuracy of each coordinate. The reliability is made of three values – one for each of the X, Y, and Z coordinate values, although there are few points with Z values. In addition to the reliability, there is information about the misclosure of

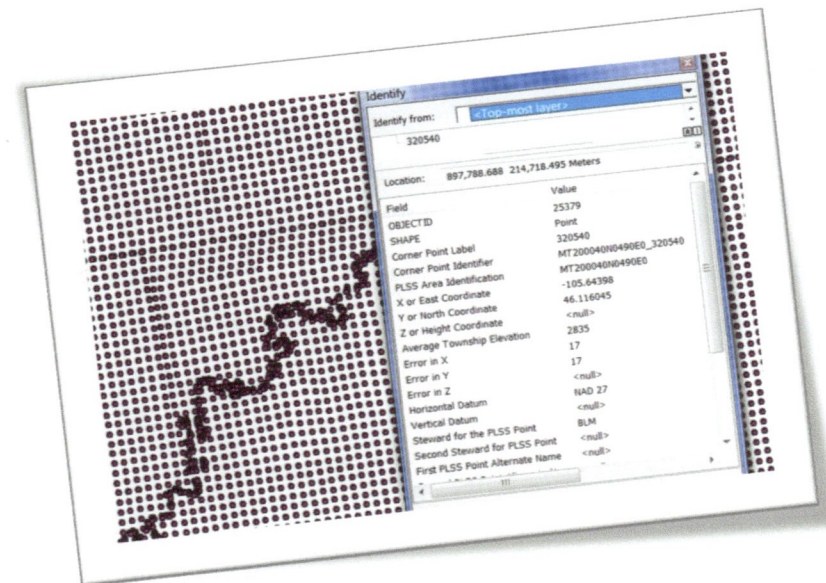

Figure 5 GCDB error estimates as attributes of the point data.

the data set as well. It is also important to note that the BLM states, "[the GCDB] …*is intended for mapping purposes only. The GCDB data served from LSI is not a substitute for a legal land survey.*" The Metadata for the GCDB contains the following statement regarding accuracy of the GCDB: *Accuracy of the individual points contained within the GCDB layers of LSI that were determined using GMM software were adjusted using both compass rule and a least squares analysis, which examines the geometry of PLSS parcels in relation to the coordinate values of known locations for control points within the PLSS grid. Each individual point carries a reliability factor indicating the error ellipse in both northing and easting which is reported after the least squares analysis is completed.*

Accuracy of the individual points contained within the GCDB layers of LSI that were determined using PCCS software were adjusted using a succession of compass rule adjustments between the control points followed by a least squares analysis, which examines the geometry of PLSS parcels in relation to the coordinate values of known locations for control points within the PLSS grid. Each individual point carries positional reliability factors for the average of the misclosures in the data set and the maximum misclosure in the data set."

The simplest way to determine the estimated spatial accuracy of a GCDB coordinate is to examine its easting and northing errors as reported in the GCDB point file. It can also be helpful to visualize the errors graphically in order to see the bigger picture across a section or a township. One may also wish to analyze a section's or a township's error statistics in order to plan a course for improving the spatial accuracy of the GCDB.

Visualizing GCDB Error

Mapping GCDB error estimates helps one to visual the magnitude and the spatial distribution of errors. There a few ways to generate maps of GCDB errors in GIS by taking advantage of the power of GIS to symbolize features based on attributes in the database. One way to do this is to load the GCDB points into GIS then symbolize those points based on the xacc or yacc values. The xACC and yACC are usually similar but they could be significantly different based on how reliable the measurements were going east-west versus north-south. It is probably more useful to use the larger of the values if there is a discernible trend within the section or township.

Point symbology based on error estimate

Dot symbols for the GCDB points may be classified by using values in the attribute table, The dots may be sized according to the magnitude of the error, or one may apply a color gradient based on a preferred range of error classes such at 0-10; 11-40; 41-80; >100.

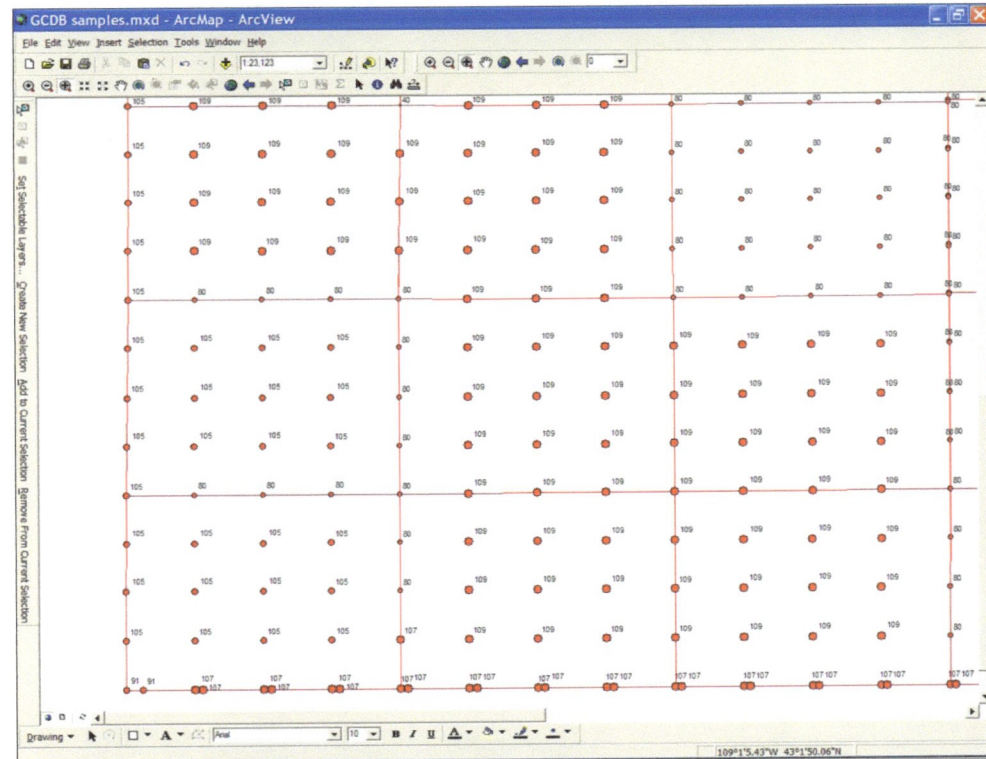

Figure 6 GCDB points symbolized by graduated size based on magnitude of error. Labels indicate error estimates (in feet)

One may also combine these methods (using size and color), which helps further clarify the distinctions as well as making those distinctions easier for those who may be color blind.

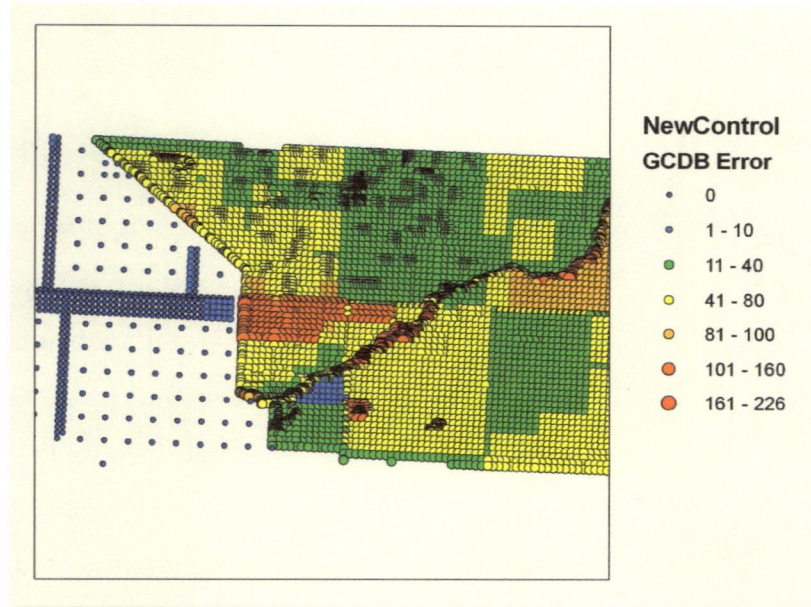

Figure 7 GCDB points colored by range of error

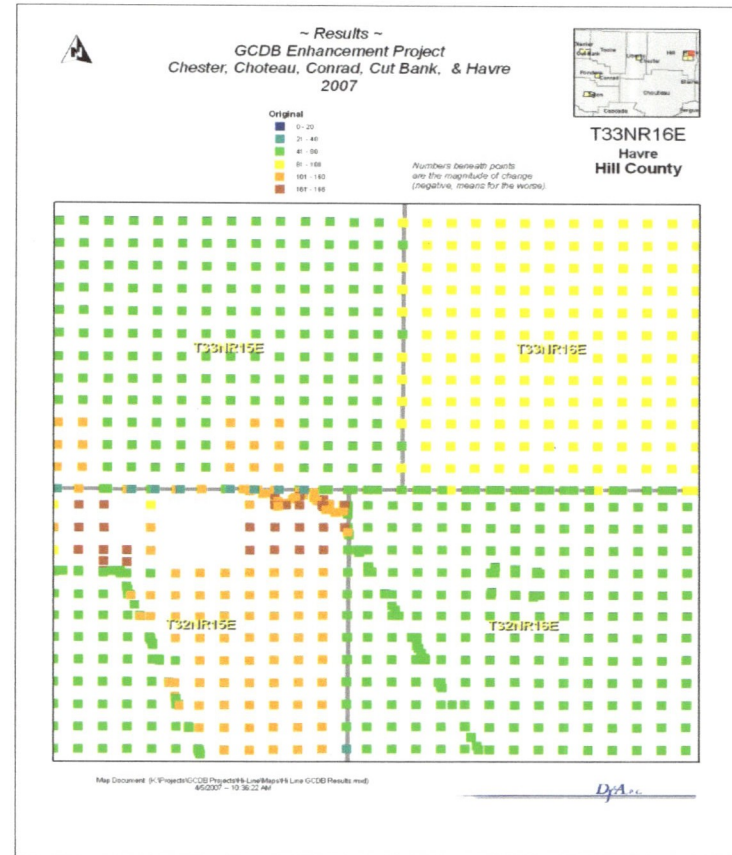

Figure 8 GCDB points symbolized by color ranges of error

Polygon symbology based on aggregated values

Point error estimates can be summarized for entire sections or townships in order to visualize errors over larger geographic areas. This is particularly helpful when analyzing GCDB reliabilities with respect to other data which depend on the GCDB - such as parcel densities.

Useful ways to summarize errors for townships are based on statistical analysis such as average error, total error, or average xACC vs. average yACC.

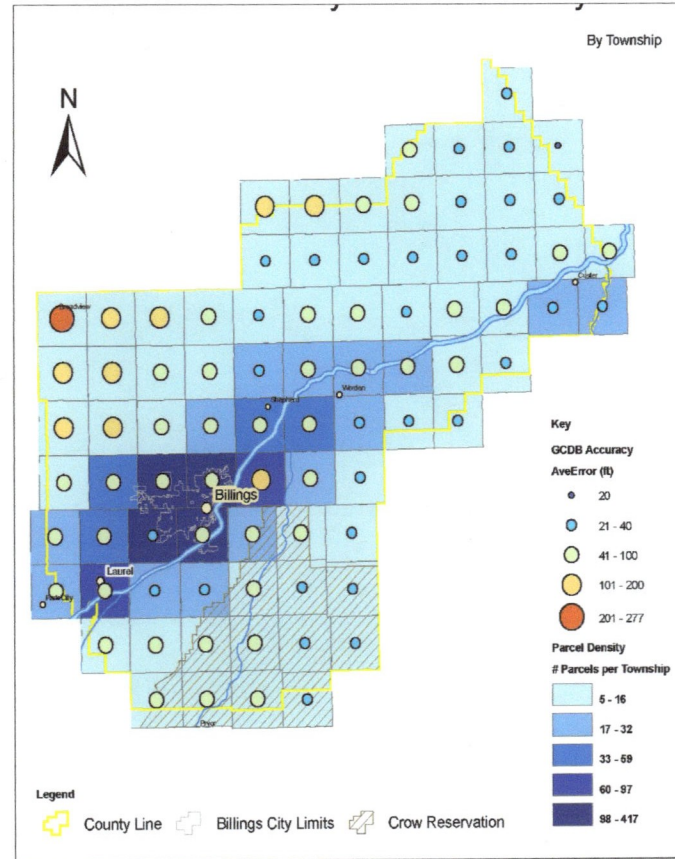

Figure 9 GCDB point values aggregated per township then the values added as color and size dots to the township polygon, shown here overlain on a parcel density calculation (blue polygons).

Chapter 2

Ch. 2 Using the GCDB in GIS

In this chapter, we discuss some of the ways to use the GCDB in a Geographic Information System for aligning spatial data and creating new GIS datasets, we also discuss some GCDB cartography.

GCDB for Mapping GIS Data

The GCDB can be used a few different ways to map GIS data. A person can geo-register GIS data to the GCDB in order to align GIS data to a standardized framework GIS dataset; or one can copy the GCDB points, lines, or polygons geometries to construct new GIS features that are based on PLSS features; or one can trace the GCDB geometry in order to conform new GIS data to the GCDB points and lines. Any of these methods will result in GIS data that conforms to the GCDB positions for PLSS related features.

Four basic approaches to mapping GIS data with the GCDB are,

    i.   Geo-register GIS data to the GCDB;

    ii.   Snap GIS feature geometry to the GCDB when creating new GIS features;

    iii.   Copy the GCDB geometry (points, lines, or polygons), to new GIS datasets;

    iv.   Move existing GIS datasets to the GCDB (rubber sheeting process)

i. How to geo-register GIS data to the GCDB

The GCDB is a reliable resource for geo-registering maps, aerial photography, and other GIS data that have points or lines based on the PLSS. To digitize hardcopy maps in order to create GIS data from them, or to display the maps as background information, the hardcopy maps can be scanned to an image format that one can then geo-register by using points on the scanned image to the corresponding GCDB points. Many maps of the western US will show some information on them that one can relate to the PLSS. The reference may be lines that follow the PLSS, such as fence lines, or tic marks that show PLSS corners, or a bearing and distance to a PLSS corner.

Figure 10 Example map as image file.

EXAMPLE Geo-register map to GCDB – historic coal mine maps

Historic maps may show PLSS features such as section corners or ties to PLSS corners, or as in this example, township lines and sections lines are drafted directly on the map. The PLSS lines on the map provide a good set of reference because of the number of PLSS intersections they produce which are PLSS corners that can be registered to GCDB points. The best practice for geo-registering data is to surround the area of interest on the map with registration points – preferably those that are nearest to the features on the map. Use at least four registration points although more is usually better – especially when some are inside the map boundary as well. Note that, because

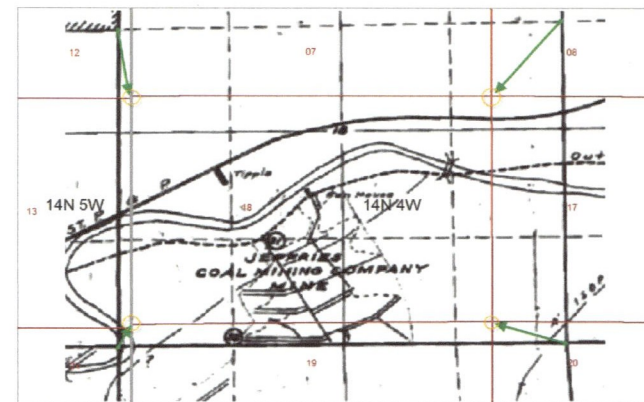

Figure 11 Geo-register image file to GCDB points using the PLSS corners shown in the map.

of errors built into the map during its drafting, some registration points might actually distort the map quite badly. Any points that worsen the map alignment should be removed from the geo-registration point set and the transformation rerun to recalculate the transformation.

EXAMPLE Geo-register photograph to GCDB – aerial photograph showing fence lines that follow the PLSS.

To be able to geo-register aerial photography one must be able to see features on the map that one can relate to the PLSS. In the example shown here, there are roads and fence lines that follow PLSS lines. The road runs north along the section line then turns east, but at that turn, a fence line continues north along the section line. One can also see a fence line running west from the turn.

The section on the left side of the photo (section 14) was apparently subdivided into 160-acre tracts based on splitting the section into quarters. The point where the road turns east may likely be the ¼ quarter. So that point could be geo-registered to the GCDB one-quarter corner (green diamond in the graphic). Following this same methodology multiple registration points can be found on the photo and used to geo-register and rubber sheet (conform) the aerial photo to geographic space so that it will line up with other GIS data. There will be some error in the alignment due to the distortion inherent in the photography, the error in the GCDB, the error in the geo-registration (digitizing points on the screen), and misalignment of other data sets.

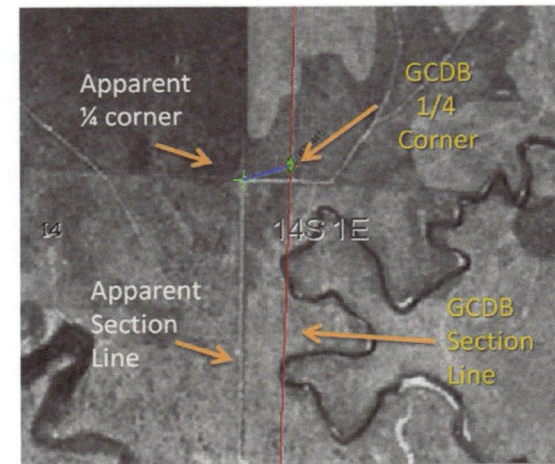

Figure 12 Scanned aerial photograph geo-registration to GCDB points.

ii. How to snap new GIS data to the GCDB geometry.

Mapping to GCDB Points

To map new GIS point, line, or polygon data to the PLSS one should *snap* the GIS geometry (points, line endpoints or polygon vertices) to the GCDB points where appropriate. For example if a GIS polygon dataset begins at a section quarter corner, one can start drawing the GIS geometry by snapping its beginning point to the GCDB quarter-corner point, then snap subsequent vertices that conform to the PLSS to their respective GCDB points (township corners, section corners, ¼ corners, and 1/16 corners.

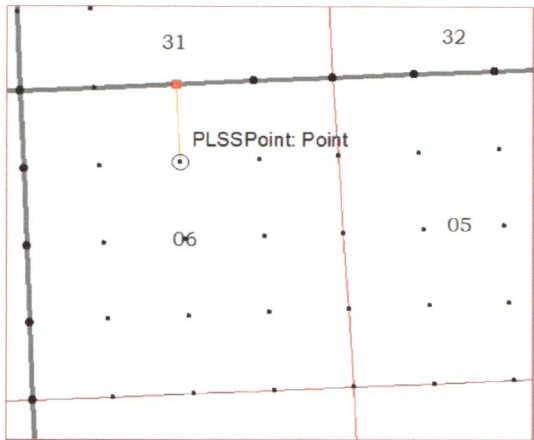

Figure 14 Snapping GIS feature vertex to GCDB point.

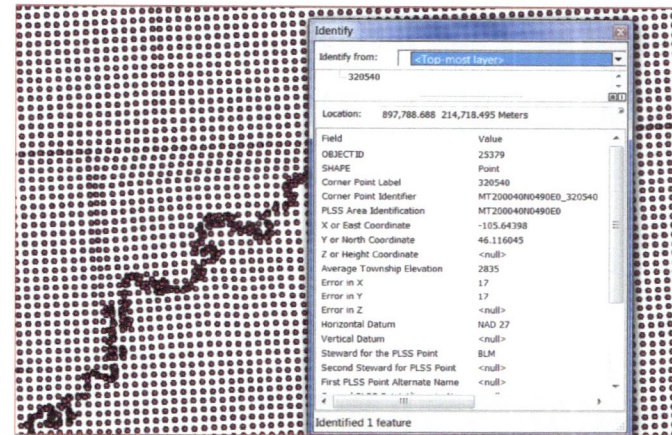

Figure 13 GCDB point attributes.

iii. How to copy GCDB geometry to create new GIS data.

New GIS data that follow the PLSS can be constructed by copying the GCDB points, lines, or polygons as appropriate in the legal description, then pasting that GCDB geometry into a new GIS dataset.

The following legal description of a rural fire district is based entirely on the PLSS. The narrative description calls for all of sections 17, 13, 22, 23, 27, and 34 in Townships 6 North, Range 6 West, and all of sections 17 and 18 in Township 6 North, Range 5 West. The GIS layer of this district can be constructed entirely by copying the appropriate section polygons from the GCDB, then pasting those polygons into the GIS layer. The attribute values and fields may have to be updated for the GIS dataset, but copying the section polygons is a very quick way to capture the complete and accurate PLSS geometry to a new GIS data set. The resultant GIS dataset of the fire district will coincide with the GCDB geometry. Creating GIS data this way also makes it easy to readjust the fire district boundary if the GCDB is ever readjusted.

Figure 15 Copying GCDB feature geometry to create a new GIS feature.

Mapping to GCDB Polygons for Aliquot Parts

You can create GIS polygon areas by copying section or township polygons from the GCDB layers then pasting those polygons into a new GIS. The pasted polygons may be merged together to create a unified polygon, if desired.

The resultant polygons can also be split by snapping a cut line to the appropriate GCDB points in order to create aliquot parts parcels. For example, to create a polygon that represent the SE ¼ of the NE ¼ of a section (the southeast quarter of the northeast quarter of a section), simply split the section into quarters by splitting the section from ¼ corner to ¼ corner, then splitting the northeast quarter again into quarters by snapping to the 1/16th corners. The resultant polygon will be the exact size, location, and dimensions of the SE NE of the section. This technique will work for any aliquot parts division down to 40 acres because the GCDB contains points for 1/16th corners for typical townships. To subdivide a parcel further, simply split the polygons by snapping to the midpoints of the polygon boundary lines where there are no GCDB points.

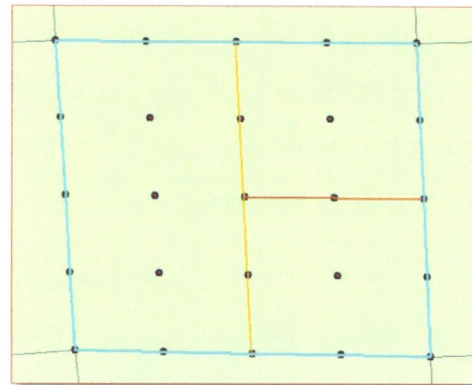

Figure 16 Splitting a GCDB section polygon by snapping a cut line to GCDB points.

Splitting a GCDB section polygon by snapping to

Case Study: Mapping Fire Districts to the GCDB

The following fire district boundary description contains elements that incorporate PLSS related features. Some of the PLSS elements include entire *areas* (e.g. all of sections X, Y, Z), and some incorporate PLSS features as boundaries (e.g. along the south line of section X ...) Because the GCDB GIS features contains areas, lines, and points it is easy to use those to accurately characterize the fire district by copying, tracing, and snapping to the GCDB GIS features. If a district describes entire sections, one can copy the GCDB polygons for those sections then paste them into the boundary feature class.

Portions of land located in Township 10 North, Range 4 .
Township 10 North, Range 5 West; Township 11 North, Range
West; and Township 11 North, Range 5 West, M.P.M., Lewis a..
Clark County, Montana, and more specifically described as
follows:

Lot 3, the S 1/2 NW 1/4, and N 1/2 SW 1/4 of Section 4; and
that portion situated North of the Southwestern R/W of the
Burlington Northern Railroad in Section 5; Township 10 North
Range 4 West.

Lots 1, 2, 3, and 4, and SE 1/4 SW 1/4 in Section 18; the
W 1/2, SE 1/4, and SW 1/4 NE 1/4 in Section 19; SW 1/4 in
Section 29; Sections 30, 31, and 32, and the E 1/2 SW 1/4
in Section 33; township 11 North, Range 4 West.

Sections 10, 11, 12, 13, 14, 15, 22, 23, 24, 25, 26, 27, 34,
35, and 36; Township 11 North, Range 5 West.

That portion of Section 1 situated North of the Southerly
boundary of Lot No. 38, M.S. No. 85; Sections 2,3,4, and 5;
N 1/2 NE 1/4 and N 1/2 S 1/2 NE 1/4 in Section 8; N 1/2 in
·ction 9; N 1/2 in Section 10; N 1/2 and NE 1/4 SE 1/4 in
·ion 11; and SW 1/4 NW 1/4 in Section 12; Township 10
. Range 5 West.

Figure 17 Example legal description of a boundary with PLSS elements.

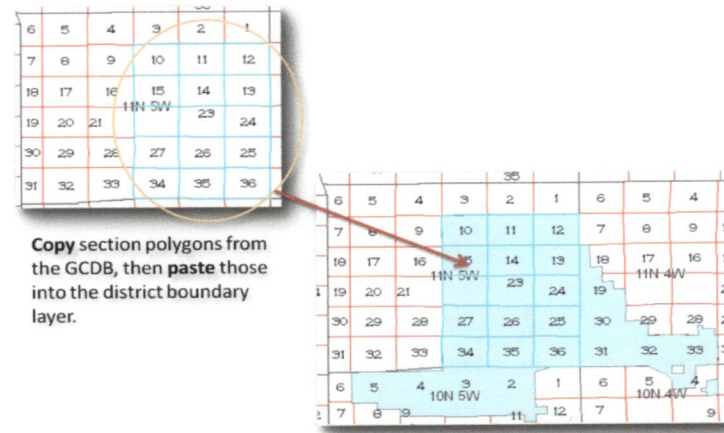

Copy section polygons from the GCDB, then **paste** those into the district boundary layer.

Figure 18 Copying GCDB geometry to paste into a GIS dataset.

If the description calls out parts of a section based on aliquot parts, then the section polygon should be split by snapping to the appropriate mid-points of the polygon boundaries. Depending on the extent of the division, a section polygon might have to be split into smaller and smaller subdivisions e.g. NE ¼ SE ¼ in section 11 requires that section 11 be split into quarter-sections, then the southeast one-quarter-section be split into quarters again to get the ¼ - ¼ section part as shown below.

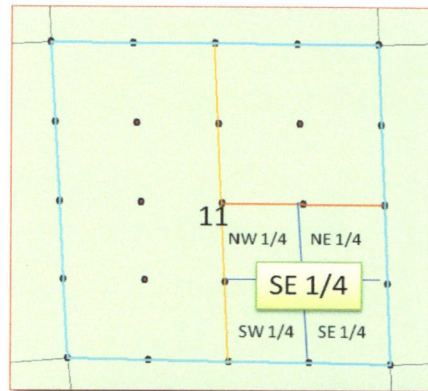

Figure 19 Subdivision of a GCDB section polygon into aliquot parts by drawing cut lines that snap to GCDB points.

iv. How to Move Existing GIS Data to the GCDB

Moving existing GIS data sets to the GCDB is a coordinate transformation process. The GIS data that were based on a non-GCDB PLSS reference can be moved to the GCDB by associating the PLSS vertices to the GCDB points. This can be done for every PLSS vertex for which there is a corresponding GCDB point. The process can be done manually by selecting each individual PLLS point, and then associating it to a GCDB. For large data sets with a lot of PLSS vertices, on can use the Cadastral Adjustment Script to automate the majority of the adjustment from the PLSS to the GCDB. In order to use the Cadastral Adjustment Script you need a control point dataset for the *OldControl* and a control point dataset for the *NewControl*, which define the displacement vectors for moving the GIS data. Since the OldControl point dataset does not exist it must be created from the vertices GIS dataset. The GCDB points define the NewControl. For more detail on using the Cadastral Adjustment Script see the section on Adjusting GIS Data to the GCDB in this book.

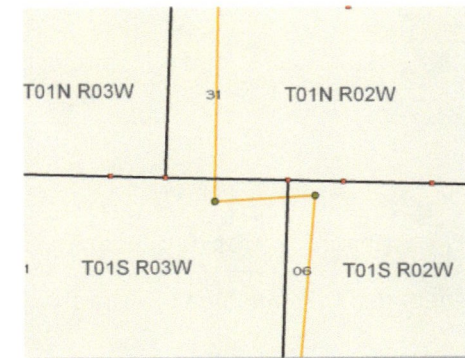

Figure 20 GIS polygon boundary (orange lines) that must snap to the GCDB (black lines). GIS vertices are green dots, and the GCDB points are red dots.

Process:

i.   Generate points from the vertices of the GIS polygon or polyline dataset and call the new point file the "OldControl";

ii.   Assign GCDB IDs to these points by performing a spatial join of the OldControl points to the GCDB point nearest them (that they *should* match to);

iii.   Run the Cadastral Adjustment Script with the GIS layer vertices (with GCDB IDS) as the OldControl and the GCDB points as the "NewControl".

The script will adjust the GIS data in proportion to the displacement vectors from the OldControl to the NewControl (within the tolerance of the search distance parameter that you set in the script).

Spatial Issues When Referencing GIS Data to the GCDB

GIS datasets may be based on the PLSS (and other features). There are many versions of the PLSS, though, for the most part, none are as accurate as the BLM's version – the GCDB. In addition, the GCDB is the official version of the GCDB (as endorsed by the Western Governors Association).

Non-GCDB versions of the PLSS are usually not coincident with the GCDB, as shown in the figure 21 below. Typically, this is because early versions of the PLSS were hand digitized from various map sources, such as the USGS topographical map series. Because PLSS datasets rarely align well with the GCDB, any GIS boundary data that is controlled by a non-GCDB PLSS will also not align with the GCDB. Figure 21 below shows an example GIS data set that does not align with the GCDB because it was originally mapped to a non-GCDB PLSS data set. This map shows a boundary layer (orange lines) that was registered to a 1993 (non GCDB) PLSS layer. It does not align with the GCDB (black and gray lines).

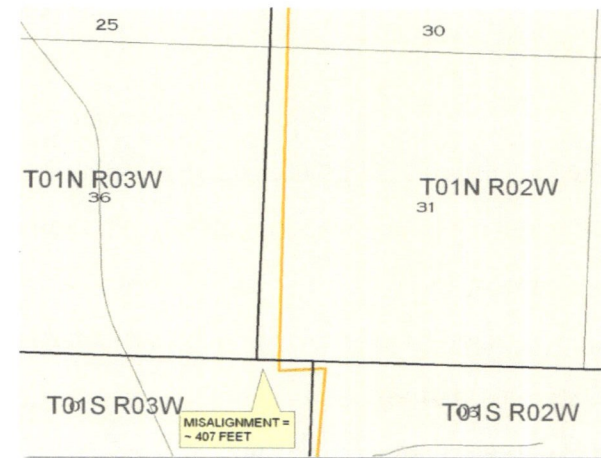

Figure 21 Misalignment between the PLSS based boundary elements (orange) of the GIS data and the GCDB (black).

Meeting the Challenges of Conforming GIS Data to the GCDB

The challenges of integrating PLSS based datasets with the GCDB can be met by performing a two-dimension ordinary least-squares transformation (aka rubber-sheeting) process. This process uses coordinate values for a before (old) control data set, and coordinate values for an after (new) control data set, to calculate the change vectors (magnitude and direction of changes) by which the GIS data set is transformed. This type of adjustment will alter the geometry of the GIS datasets by varying magnitudes and directions based on each GIS vertices' proximity to the change vectors. (Note: GIS *vertices* are the angle points for the GIS lines and polygons).

An additional consideration when performing the transformation of a GIS dataset to the GCDB or to a re-adjusted GCDB, is that many boundary layers contain non-PLSS references, such as rivers, mountain ridges, roadways, etc. which are not represented in the GCDB. The non-PLSS boundary segments typically should *not* move, when adjusted to the GCDB or re-adjusted to a new GCDB. Moving a boundary layer, such as a fire district boundary to the GCDB may require freezing, or holding fixed in place, the non-PLSS segments. Holding these non-PLSS segments fixed in place, presents its own set of challenges when using any sort of automated transformation process. Here we outline some recommended procedures for dealing with these challenges in the hope of simplifying and standardizing the process to make GIS data match the GCDB.

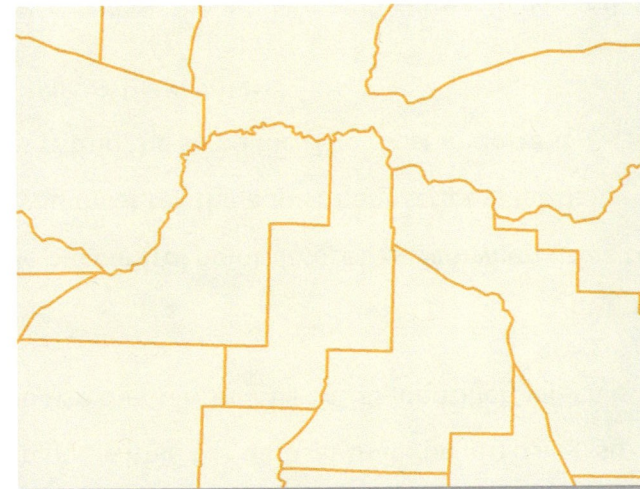

Figure 22 Boundary with PLSS and non-PLSS elements.

A Note about Point identifiers

Point identifiers (IDs) can differentiate between GCDB points, using the standard GCDB IDs, and non-PLSS that may use any form of identifier. All the IDS must be unique within a file (Old Control point file or the New Control point file), but any IDs that are identical in the Old and the New point files, will be used to generate the displacement vectors. When using a GCBD file, you can use the unique point identifier already in the GCDB file. Note that the identifier must be unique within the dataset that you are using, so although you could use few different fields within the GCDB, it is normally safest to use the fully unique GCDB identifier, which is unique within a state and, across states.

You can also invent point identifiers for both the new control point dataset and the old control point dataset – even when using the GCDB ID. The field name for the point identifier may be called anything, although you must tell the adjustment script what that name is. You may even mix point IDs that you invent with the real GCDB identifiers within the same file. The most important consideration for the point identifiers is that each record within a file must have a *unique* (its own) ID.

The IDs in the old control file and the new control file that are the same are used to calculate the adjustment vectors. In general, the solution to fixing the non-PLSS positions so that they do not move, is to create two sets of control points for them that have duplicate IDs and duplicate coordinates in the transformation control point files. Because the rubber-sheeting method does not permit weighting, the recommended method for constraining certain areas or features to their original position, that is fixing their positions, is to ensure that the points have the same coordinates in the New control file as they do in the Old control file. One way to do this is to simply copy their IDs and coordinates from the Old control to the New control. The process of fixing positions can be applied to any feature or parts of features that are based on mapping controls that are superior in accuracy to the GCDB point positions.

GCDB Cartography

The GCDB data set contains sufficient geometry (points, lines, and polygons), and attributes for a variety of cartographic purposes. The data sets contain polygons for townships, polygons for sections, polygons for label descriptions (e.g. NWNW), points for township, section, and section-breakdown corners and other points.

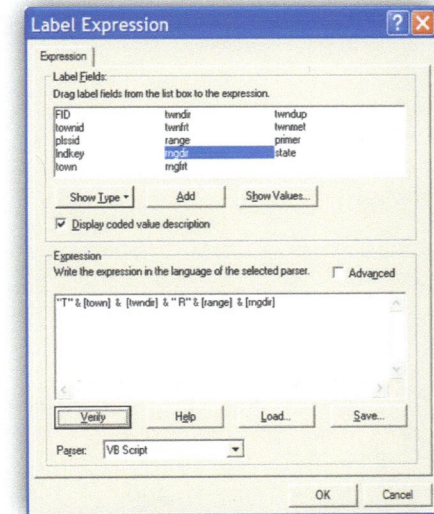

Figure 23 Formatting GCDB attributes for labeling.

The rich attribution of the GCDB data allow for a variety of labeling and cartographic options, although some of the labeling requires a bit of string manipulation to conform to standard output. For instance, to create a typical township label requires concatenating four different fields along with some additional text, as shown in the figure 23 Formatting *a Township String*. However, the cartographic results are quite workable as can be seen in the *Cartography Example*. The township and section lines shown in the cartography example were created from the two polygon files. In this case, only the boundaries of the section and township polygons are shown. The labeling is based on the township attributes of the township polygons, and the section attributes of the section polygons. In this particular instance, the GCDB data are stored in a geodatabase, which provides better speed as well as data integration.

Reliability Cartography

One cartographic product that surveyors would find of interest is the reliability diagram. One can generate a diagram of the reliability of the coordinates for each PLSS corner in the GCDB from the accuracy values in the COORD point file (coordinated points). The COORD point file has accuracy fields for the X component and the Y component of the horizontal accuracy of the coordinates for each point. The values given are in feet. Typical errors are around 40 feet, but as shown in the reliability graphic, larger errors are also possible. The magnitude of the errors is a function of the amount and quality of control and survey data that went into creating the GCDB for any particular area. Some areas had lots of survey control, and good supporting survey data. Other areas had poor surveys and/or little survey control. Therefore, the results vary considerably. Anyone who is not satisfied with the accuracy of the GCDB in their area of interest can work with the BLM to reduce the magnitude of the errors. The BLM is drafting a process for improving the spatial accuracy of the GCDB using a combination of inputting

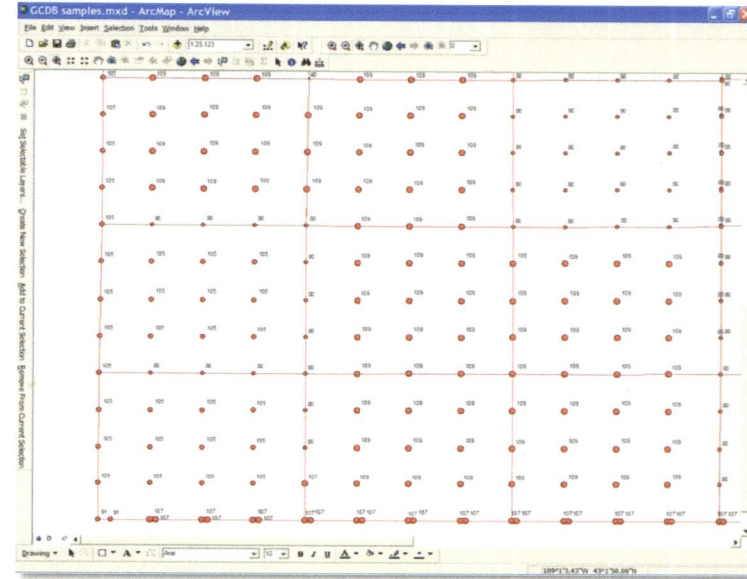

Figure 24 Labeling GCDB points with error estimate values (feet).

existing survey data and acquiring new survey control. This process will be more fully described in a later issue. The overall horizontal reliability of the coordinate can be calculated from X and Y error estimate vales, and then symbolized, such as shown in the *Reliability of Coordinates* graphic shown here. In this example the size of the dot is related to the magnitude of the estimated error, the larger the dot, the greater the error. The numbers beside the dot are the magnitude of the estimated error in feet. This method is a quick and straightforward characterization of the distribution of error across this particular township. Perhaps more useful is to calculate the reliability of entire sections of lines between points, in order to help understand the spatial distribution of the errors. Such calculation and conflation of values from the point geometry to the polygon geometry requires some GIS analysis work and a little computer programming. Nevertheless, it is useful when viewing the reliability from the perspective of polygon geometry such as the reliability of parcel boundaries based on GCDB.

The GCDB can also be used to create PLSS cartography for base maps – typically showing townships, range, and section lines (using either polylines or polygons) along with the associated annotation, as shown below. One way to do this is by inserting the GCDB polygons for Townships and for Sections, then setting the cartography and labeling.

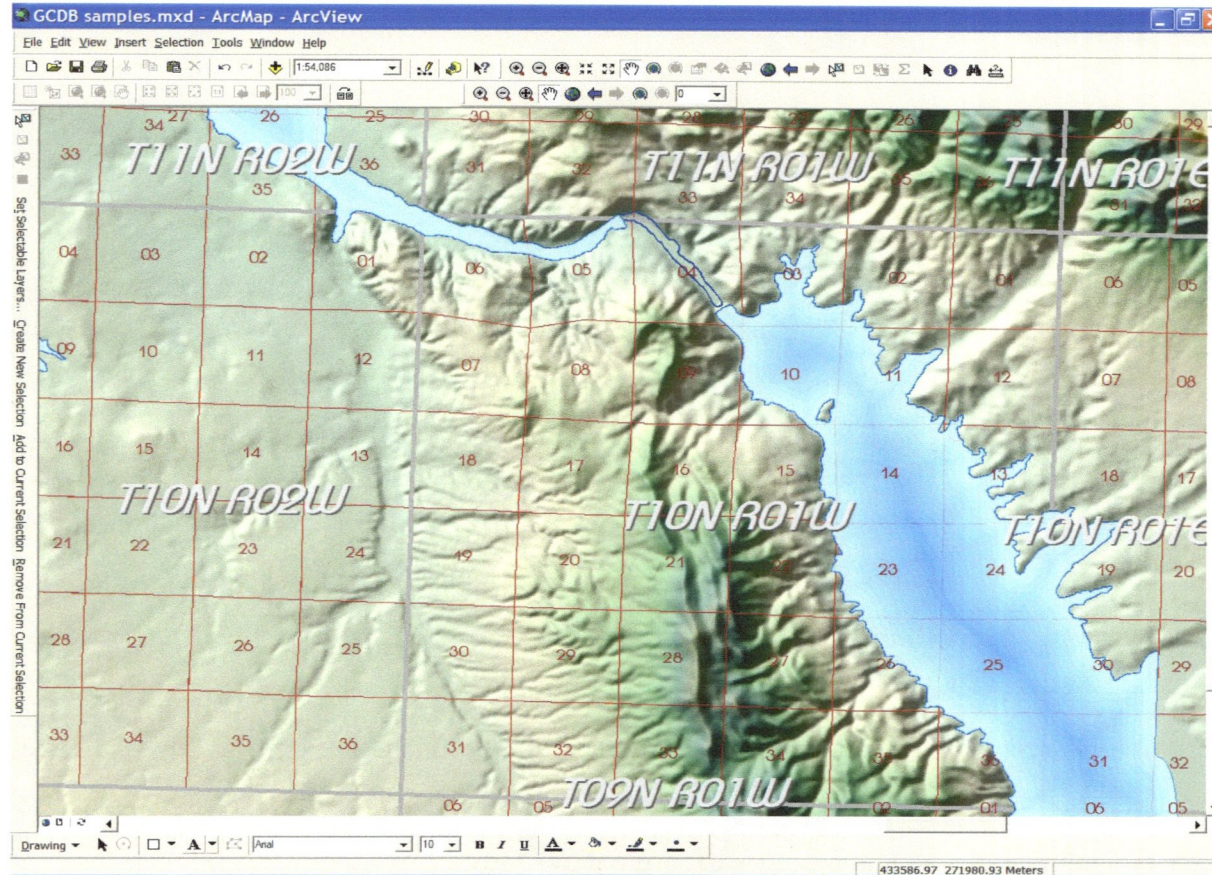

Figure 25 Example PLSS cartography using GCDB features.

Chapter 3

Ch. 3 Using the GCDB in Surveying

In this chapter, we discuss some of the ways that the GCDB helps surveyors' field operations and using the GCDB to organize survey information in the office.

Introduction

The GCDB is a coordinate framework for the PLSS and thus is a set of coordinate values for each township corner, section corner, quarter-section corner and other corners of the PLSS. Since legal descriptions for most of the western United States originate from the PLSS, the GCDB is an important component of the cadastral (land ownership) layer in the American west. Surveyors can use the GCDB coordinates to search for corner monuments in the field. The GCDB can also serve as a mapping basis for creating a graphic index to survey records.

PLSS and GIS

The PLSS is used in GIS in the west for creating parcel layers as well as other GIS layers. Since legal descriptions, surveys, and plats include ties (bearings and distance) to the PLSS, the PLSS is the framework of control upon which the parcels (ownership boundaries) and many other GIS layers are built. Some methods for creating the GIS layers begin by creating the PLSS layer then digitizing the parcel lines into the PLSS. Other methods may use coordinate geometry (COGO), whereby the legal descriptions are entered into the computer, such as a bearing and distance line originating from a section corner. An example of this is the Montana Department of Administration's Cadastral Mapping Project which used the GCDB framework for the department's automated mapping program. The program used by the Montana Department of Administration, created parcel polygons from the aliquot parts legal descriptions that resided in text form in the Department of Revenue's Computer Aided Mass Appraisal (CAMA) files. The GCDB served as the beginning line work from which the aliquot parts are created. For example, if a legal description for a parcel was *the NW ¼ of the NW ¼ of a section*, the program would subdivide the section as defined by the GCDB, split the section into quarter sections, then split the NW ¼ into quarter-quarter sections. The NW ¼ of the NW ¼ then becomes the parcel polygon for that legal description. The result was a GIS feature of the ownership which can be queried for ownership, land values, improvements and other data in the

Department of Revenue database. The GCDB provided a convenient and consistent framework for parcel creation across all 56 counties of Montana.

Other Uses of the PLSS

Since the PLSS is a common frame of reference in the west it is also often depicted on kinds of maps other than cadastral maps. Therefore, the PLSS is frequently used as a reference for locations of other GIS features. For example, a water source may be referenced to a section corner in a legal description. Additionally, search and rescue operations rely heavily on the PLSS as reference for location and logistics. Additionally, surveyors can use the coordinate data of the GCDB to perform field searches for PLSS corners. To find a corner a surveyor could obtain the coordinates from the GCDB, enter those coordinates into a GPS unit, and then navigate to the search area. Since the GCDB is derived from the GLO (and in some cases more recent survey data), the GCDB makes corner searches more efficient.

PLSS Accuracy

The higher the accuracy of the PLSS, the more accurate will be the GIS layers that are built upon it. However, there are various sources for a PLSS layer and each source has its own accuracy. One of the most commonly used sources has been the U.S. Census Bureau's TIGER files. The TIGER files are digital layers (including roads, railroads, rivers, etc.) that were digitized from the 1:100,000 scale USGS topographic maps. The resulting accuracy of these data is ±170 feet. Although this scale is appropriate for small-scale mapping over large areas such as an entire state, it is inappropriate for large scale uses such as those used by land surveyors and local governments. Other sources are hand digitized PLSS from 1:24,000 scale USGS TOPO maps to improve the accuracy.

The corner positions of digitized work are inherently less accurate than coordinate geometry-based positions. Using different sources, scales and methods to create the PLSS layer causes conflicts among the resultant data sets, because each method will derive a different coordinate for the same corner. For example a hand digitized PLSS from a 1:24000 scale map, would poorly edge-match a TIGER PLSS layer. Therefore, it is important to use a single source for the PLSS, which of course is the GCDB.

GCDB for Corner Searches

Surveyors can obtain the corner coordinates from the GCDB to input into a GPS as an aid to searching for corners in the field. The GCDB provides coordinates and error estimates of those coordinates (easting error and northing error). Depending on the capability of your GPS, you may need to hand enter the coordinates to your GPS or you may be able to upload a file of the GCDB coordinates to your GPS.

Figure 26 PLSS corner monument brass cap.

Methods

1. Hand enter the GCDB coordinates into the GPS, then use the GPS unit's Go To or Navigate (to) options to generate a distance and direction to the corner from the current GPS location.

2. Download the GCDB coordinates to a file, convert the coordinates file to a format that your GPS can use, load the GCDB data into your GPS. Depending on the type of GPS equipment you have, you may see the GCDB points on the display (possibly with other background map data), and you may be able to see the GCDB information such as the point ID and coordinate reliability. Most importantly, you should now be able to navigate to the PLSS points based on the coordinates that are now loaded into your GPS.

GPS navigation to GCDB point

GCDB point IDs
info on GPS

GCDB points shown
on GPS

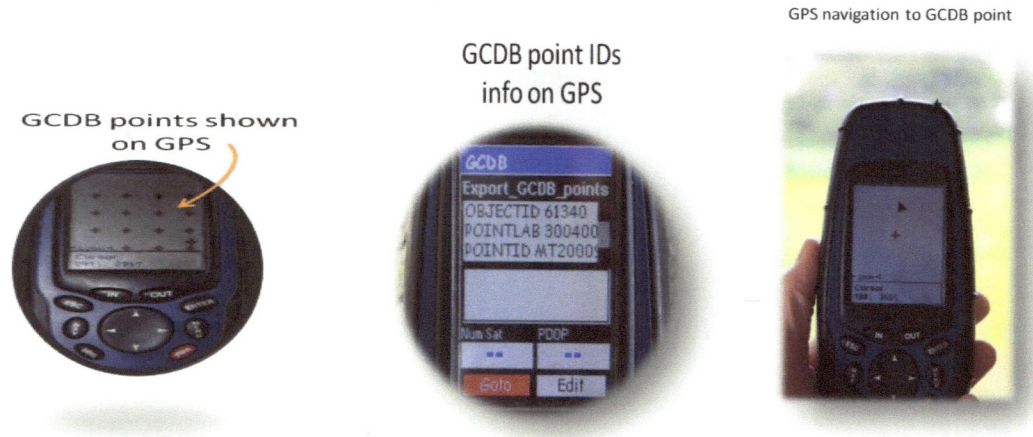

Figure 27 Handheld GPS with GCDB points loaded for finding PLSS corners.

Map Road Rights of Way

Public road rights of way (ROW) are importnat to surveyors, road maintentance departments, construction engineers and planners, natural resource managers, and others. However, access to public right of way documents has historically been difficult because so few of the documents are digital, and because many of the filing systems are manual. Additionally, interpreting right of way documents can be challenging because the legal descriptions may not be mapped or drawn graphically but may only be described by courses and widths. Additionally, the right of way itself might not be legally, where the road is was built physically.

Creating a GIS for road rights or way can ameliorate some of these issues. A right of way GIS can facilitate discovery of ROW conveyance documents, quicken access to the documents, can show where the right of way is located, and could even show the exact geometry (courses and widths) of the right of way. Once the ROW GIS data are created those data can be used on a desktop computer and overlain with other GIS data such as aerial photography, engineering designs, zoning, terrain, and others. Additionally, the ROW GIS may be served as a resource on the internet in a web mappng application (see Minnesota's ROW web application), or even as a map service that can be consumed on client desktops via the internet.

Online access to Rights-of-Way information using a map interface.

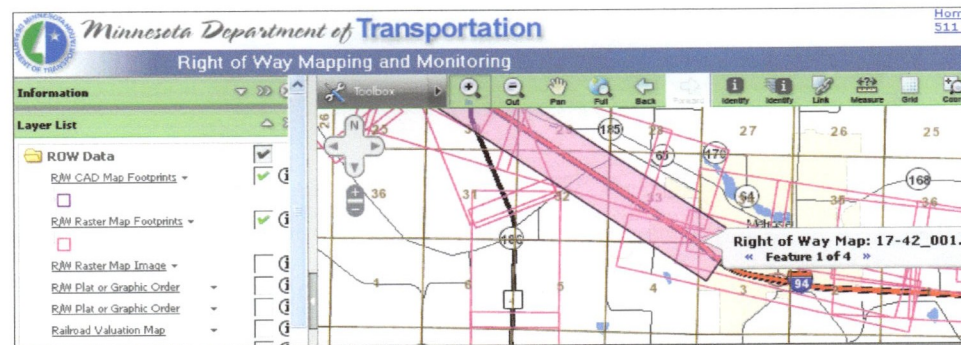

Figure 28 Example online access to right-of-way information using a map interface (Minnesota Department of Transportation).

Depending on objectives, schedule, and cost, you could use one of three methods to create a road right of way GIS layer. I categorize the three methods discussed here by the GIS geometry used to represent the location of the right of way project. The method you use will be driven by how useful the product must be for you, how much time you have to create the product, and the cost to do the mapping. The choices of GIS geometries are Point, Line, and Polygon any of which you may use to create a GIS index for the right of way.

Geometry	Speed	Accuracy	Link to Document	Spatial Analysis
Point	Quickest	Least	Yes	Count, approximate near, approximate distance
Line	Moderate	Moderately	Yes	Count, near, distance, intersect
Polygon	Slowest	Most	Yes	Count, near, distance, intersect, contains

Point

The simplest and quickest way to map a right of way project is to create a GIS point for the location of the project, then associate the right of way project documents to the point. The associated documents could be scanned images of plans, or CAD drawings, PDFs of conveyance documents, etc. The project documents may be stored independently of the right of way project GIS data and a table or database would store the linkages between the two sets of data. This can be done as attributes in the GIS dataset that contain the file name to the documents, or a separate table that relates the records in the GIS file to the record or records in the documents database.

The advantages to creating a point-based ROW GIS are that they can be created very quickly and easily. The points can represent the middle of the project, or the starting point or end of the project. If the ROW has references or survey ties to the PLSS then using the GCDB points helps to get the project point in the correct location. GCDB points can be used to map road right of way projects in a couple ways depending on whether the project has survey ties to the PLSS or not. If the project does have ties to the PLSS then the corresponding GCDB point(s) may be used.

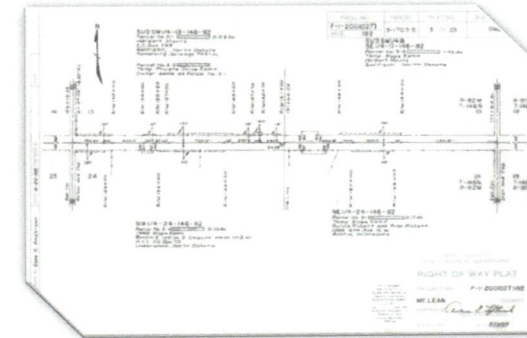

Figure 29 Scanned image of right-of-way plan drawing.

The disadvantage to using point geometry for a GIS index of right of way is that a point will not represent the full extent, length, width and geometry of the right of way. For display purposes, a point will not show what the right of way looks likes. Additionally the ability to perform spatial analysis is greatly diminished. For example: determining whether a certain condition (such

System Architecture for Publishing ROW Information

as wildlife habitat) falls within the right of way cannot be done reliably with a point. However, it is possible to perform buffer analysis using a point, such as whether or not a condition is within a certain distance of a project center or end. Usually, spatial analysis of a point which represents something that may have a large and complex shape can only be a rough estimate, at best.

Line

Right of way projects can be mapped using a line, in much the same way as a point can. The line can represent the location and the linear extent, as perhaps the centerline geometry of a right of way project, which can also be linked to right of way source documents and tabular information.

Using a GIS line to create a right of way index has advantages over using a point in that the full linear extent and possibly the linear shape of the right of way can be drawn for display, which would allow improve spatial analysis. With a line, spatial analysis such as *within distance of* is improved by the ability to search from either end and anywhere along the line of the project. With the line geometry, it is easy to generate a buffer of the full width of the right of way (if the width is constant). In addition, by using a line to represent the right of way, it is possible to analyze whether something crosses the right of way or whether the right of way crosses something.

The disadvantage to mapping lines as opposed to points, is the creating the line take more time, whether the line mapped is generated by sketching or by typing in the coordinate geometry (bearings and distances of the centerline for example). However, if the geometry already exists in a digital form such as CAD, then creating the line can be as straight forward as converting CAD geometry to GIS features. Such a conversion can be done quickly and inexpensively.

Polygon

For greatest utility, a GIS index of right of way should be mapped as polygons that represent the entire right of way geometry. The most accurate method for mapping the geometry is by COGO'ing in the right of way bearings and distances. Where appropriate, references to the PLSS should use the GCDB to locate the polygon in geographic space. A polygon can best represent the complex geometry, extent, width(s), and location of the right way, and thereby provides the best visual representation, and affords the most accurate spatial analysis.

The disadvantage to creating the polygons is the time that it takes to type in the complex geometry. However, if the geometry is already represented in digital form such as a CAD drawing, then the entire process can happen far more quickly and with less likelihood of mistakes in data entry.

Alternatively, complex geometry can be generalized to a simple rectangle to show the approximate length and approximate with of the project as shown below in the Minnesota Department of Transportation right of way index map.

All three methods described above may be used independently, or all three methods (geometric representations) can be used in a phased approach – starting with the simplest geometry to generate a quick index, then creating the more complex geometries as time and budgets allow. If two or three of the geometries are available then they can also be used for scale-dependent mapping applications such as web map services. The different geometries can have scale dependencies so that, for example at small scales the points appear, at large scales the lines appear, then at very large-scale, the polygons appear. Regardless of the GIS geometry type that you use to create your ROW GIS, any tabular data that you capture can be associated with the GIS features, including hyperlinks to digital source documents.

Create a Survey Index Layer

Whether you work in a county surveyor's office or a private survey firm, you can create a GIS layer of surveys that can be used for a variety of purposes. For example, performing a search for a particular survey to see where it is located is a quick and simple task in GIS. Another example is finding the adjoining properties to a survey in order to produce a mailing list of owners *(based on a parcel layer—see example later in this section)*. Finding all the control points (from a control point layer) within one-quarter mile of the survey, or showing the juxtaposition of all the surveys near a particular property and listing them by date are other useful query and analysis tasks that can be easily and quickly performed in GIS.

The GCDB is a good resource for mapping survey data to create GIS features that maintain topological relationships. Since, in the western states, cadastral survey projects are tied to the PLSS, the GCDB is a logical resource to map the location of past and future cadastral projects. Entire townships, selected sections, aliquot parts subdivisions, particular lines, and specific corners can all be mapped using the GCDB geometry of polygons, lines, and points. You can copy, trace, or snap to GCDB geometry to create new GIS features

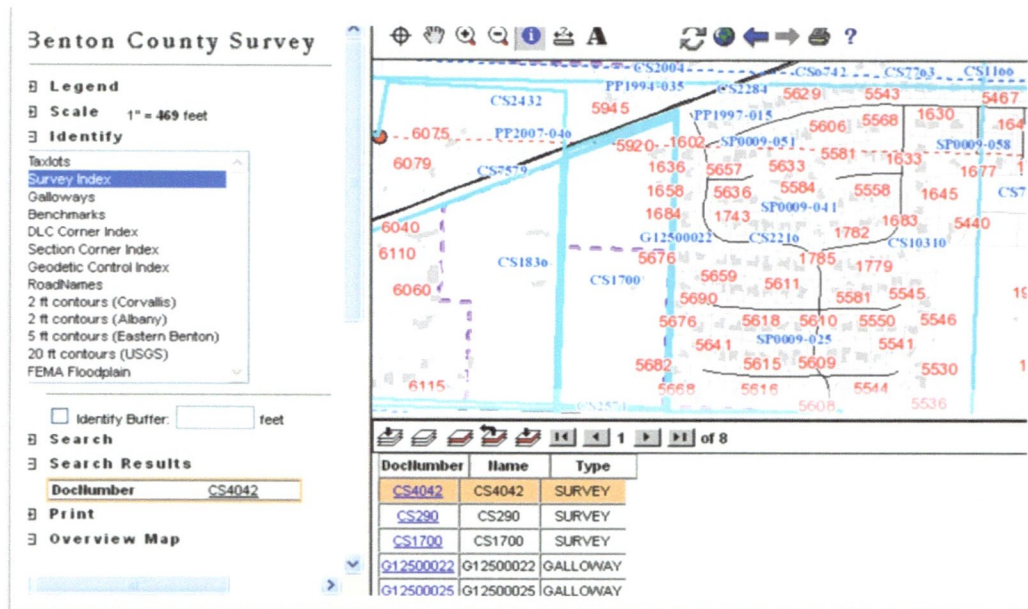

Figure 30 Example online survey index - Benton County Oregon.

and datasets to represent the locations of cadastral projects. Once you create the GIS geometry attaching associated data to it is fairly straight forward. You can create linkages to associated documents such as survey plats, scanned corner records, and project database records, or add new data as attributes to the features. Since the GCDB is delivered with topological integrity, (e.g. GCDB polygons do not have gaps or overlaps), using the GCDB to build new features starts with good data.

Similar methods to the one shown below for the Corner Records Index, and illustrated elsewhere in this handbook, may be used to create indexes for survey projects. The decision on which GCDB geometry to use for creating a GIS layer will depend on how you wish to represent the survey data. For example, to map a road right-of-way survey you would use a point (at the beginning or middle or end of the project area), or a line representing the approximate centerline location, or a polygon to represent the boundary of the enclosed right of way. Additionally the GIS geometry could be an approximate

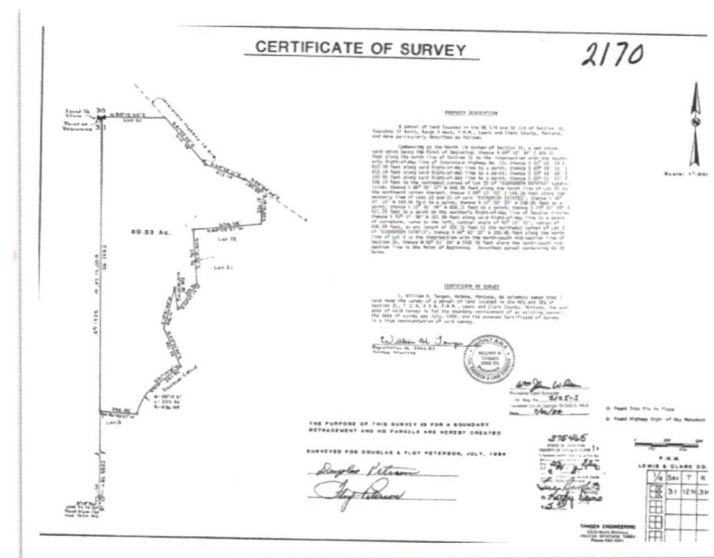

Figure 31 Example scanned image of a survey plat.

location or the exact surveyed geometry, depending on the purpose of the GIS index. If the intent is to use the GIS feature used to indicate the number and approximate location of projects and provide a pointer to the documents, then one can more quickly map them out. However if the intent is to have an exact geometric representation of the survey shape and mathematics, then more time is required to replicate the correct geometry and to geo-reference it in order to properly locate the project in the GIS coordinate system (which might be a projected one).

There are many different ways to create a survey index layer. The ideal way may be to enter the survey data using COGO with some GPS ties of the corner positions of the surveys. This is typically the most accurate method for entering the information, but it is also the most expensive and time consuming. Although COGO and GPS may provide the most accurate results, the pay-off may not be high enough to justify the effort, expense, and cost. When deciding which methods to use, consider the following factors:

• How the data will be used: the resultant accuracy should meet the needs of the intended use. Consider the kinds of queries and analysis to be performed with the data (such as listed above).

• Accuracy of the other GIS layers: for example, if the other GIS data is +/- 40 feet, then COGO and GPS accuracies of the survey index data to less than 1 foot would be overkill or may not even be achievable. Not all work needs to be done with the precision of a Swiss watch. On the other hand, if the survey layer will be used as the basis for other mapping, then the goal should be to obtain the highest feasible level of accuracy.

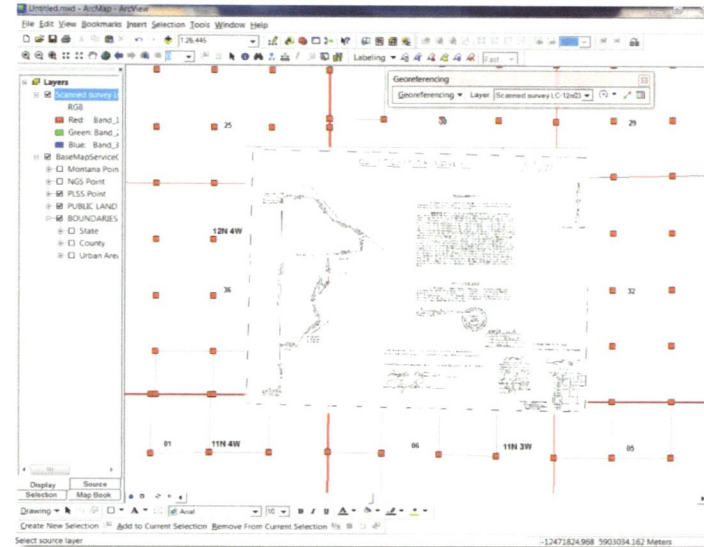

Figure 32 Scanned image of survey fit to the screen display for geo-registration.

• Budget: no one has an unlimited budget, so needs and wants must be reconciled with budget constraints.

• Time and resources to complete the work: some methods take more work and more time than other methods.

• Technical skills of staff: the skills and expertise of available staff may limit the options or expand them.

The decision on which method or methods to use may also vary depending on the level of development in the area of the surveys. For example, in urban areas, COGO combined with field ties may be necessary due to the density of surveys and the small size of parcels, streets, and other features in close proximity. Rural areas, on the other hand, tend to have large parcels and fewer surveys, so they may not require such painstaking effort. Other methods, such as described here, may be used in rural areas to obtain satisfactory products more economically and quickly. Below is a quick and very useful method that may not be viable for all situations, but can certainly achieve a high rate of production that yields good results at a low cost.

Based on the above discussion, the following are requirements for the GIS layer of surveys:

1. The GIS layer will be a vector polygon layer to allow for spatial query and analysis.

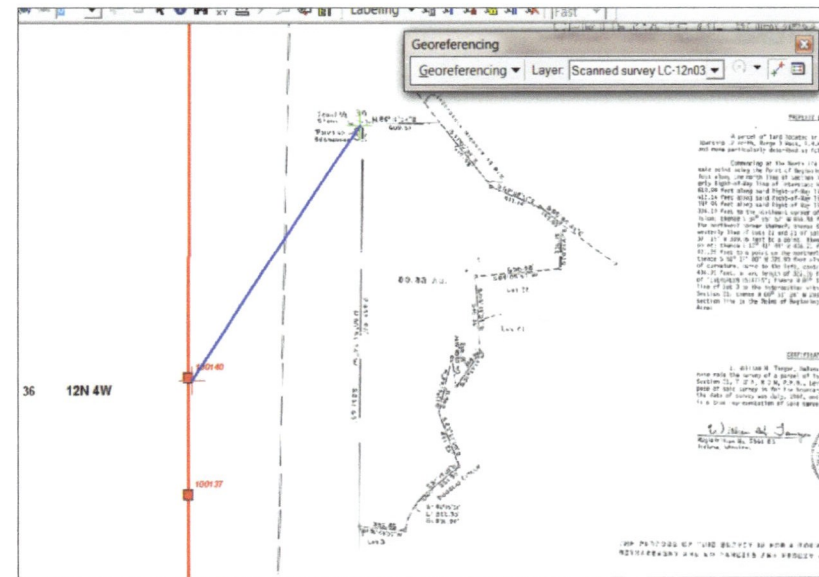

Figure 33 Geo-registering scanned survey.

2. The surveys will be represented as polygons (even if the survey itself is linear).

3. Only the bounding perimeter of the surveys—not the internal lines—will be digitized.

4. Each survey feature (polygon) will have at least two attributes: a record number (a unique ID which may be based on the public record index number or the office filing number), and a file name for the source CAD drawing or scanned image of the survey (if one

exists). Others may be added to allow other types of queries, but for this discussion only the two are necessary. It is important to note that if a survey database already exists, then the only attribute required to link the GIS feature of the survey to its record in the database is the unique ID.

5. The source documents (either CAD drawings or scanned images) will be onscreen digitized in a coordinate reference system (such as UTM, State Plane, or local datum).

Additional notes: all of the survey perimeters will be digitized, ignoring any overlaps between surveys. The monument and control points in the survey will not be captured here. Those points could be captured in a separate GIS point layer (the GCDB).

Creating the Layer

The first step is to set up an ArcMap project with the reference data. In this case, a layer of Public Lands Survey System (township, range, and section) information – the GCDB, along with the existing parcel layer will serve as the reference layers. Those layers have a spatial reference system such as State Plane or geographic coordinates, associated with them.

Next, a scanned image of a survey is loaded into the map. Initially, the survey cannot be seen because the scanned image (in TIFF format) does not have any geo-reference information associated with it; that is, the file of the scanned image has no coordinates ().

ArcMap has some geo-registration tools built-in, which can be used to quickly register the scanned image to geographic space. One of those tools allows us to fit the scanned file to the display—remember this is just a raster image (). This works very quickly and helps to get it on the screen so that corners of the survey can be picked in order to link to control points on the reference layers (GCDB points for example).

Once the raster image is visible on the screen, points on the raster of the survey can be linked to the parcel corners that they represent). As each of these registration points is linked (from image-point to control-point) the raster image registration is performed on-the-fly. Bad points can be eliminated to achieve the best possible fit. Since we are creating a survey index layer we only need to digitize the survey geometry of the scanned plat.

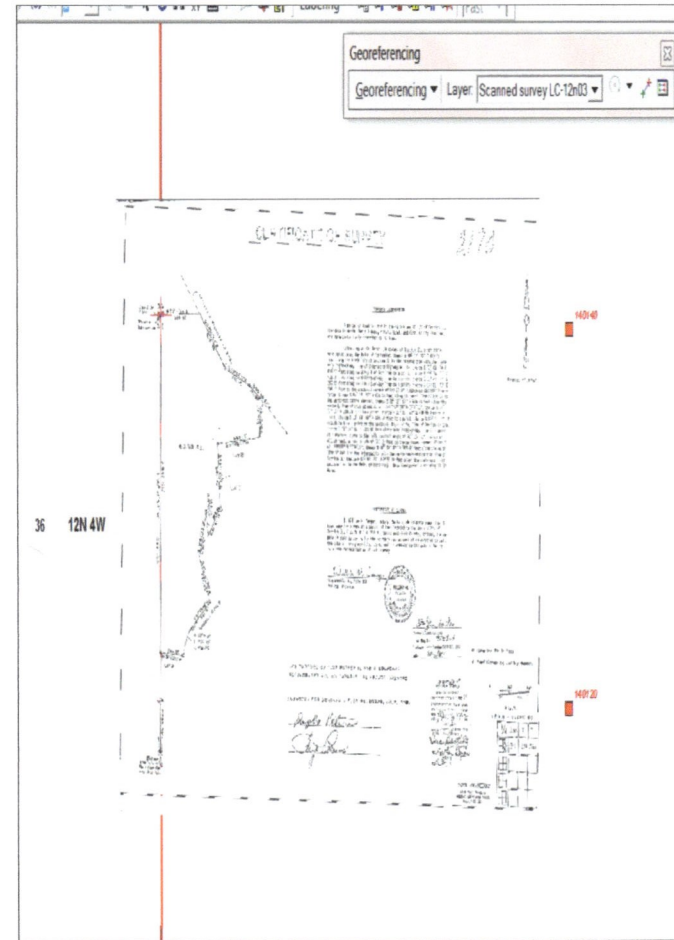

Figure 34 Scanned survey geo-registered to GCDB points.

After geo-registering the scanned image we begin editing a Surveys shapefile layer to create a GIS polygon of the survey boundary. This layer has two attributes, RecNum and FileName, for the source document reference number and the name of the PDF or JPEG or TIFF file of the scanned image. These two attributes provide a unique identifier for each GIS feature and provide a means to access the image file of the survey. After creating the survey polygon feature, the attributes may be entered.

By having this as a polygon vector GIS layer, spatial query and analysis may be performed on that layer. The file name of the scanned image of the source document has been added as an attribute to the polygon, which can be used as a hyperlink to the document image file. (A hyperlink enables the user to click on the survey polygon and display a scanned image of the original document in another window. The hyperlink to the scanned image of the source document allows the user to read the image of the survey itself in order to find any other information that it may contain. This particular survey has information about some

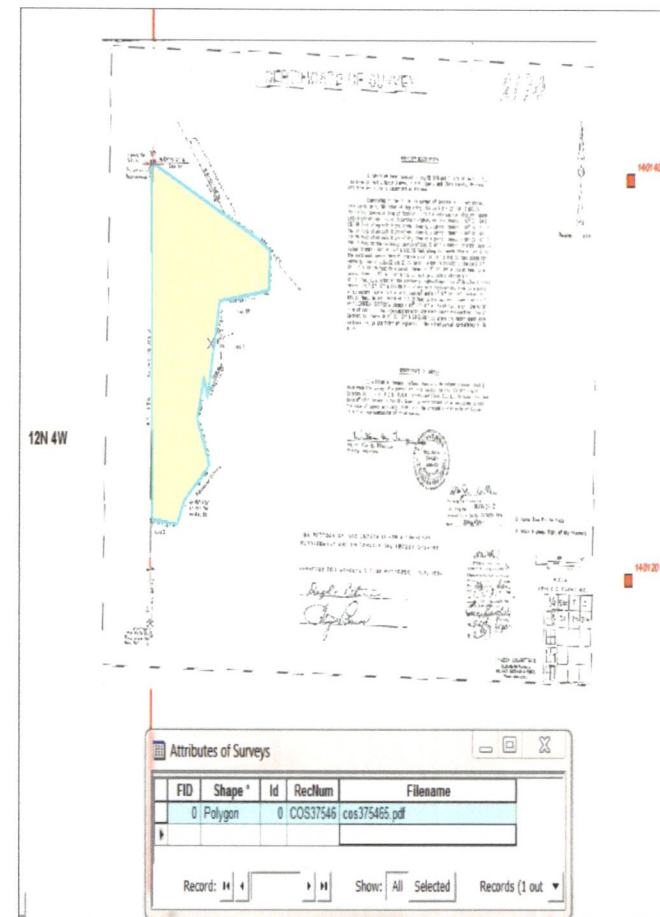

Figure 35 Digitized survey boundary as a GIS features with attributes.

of the corners, senior lots, parcel reconfiguration, and other interesting information.

Analysis

In addition to being a helpful tool for organizing, displaying, and searching for surveys, the survey index can be used for spatial analysis. GIS allows us to perform a spatial query that helps to select features of one GIS layer based on their spatial relationship to the features of another layer. We the survey index layer we can will perform a spatial query to acquire a list of owners of the properties that are adjacent to this survey for example. In addition to spatial queries, which tell us something where our surveys are with respect to other things, we can analyze the surveys themselves to obtain insights about when surveys were done, or the average size of certain types of projects, where the most recent project were done, which neighborhoods have the most work going on, etc.

Mapping Corner Records to the GCDB

Some western states require that surveyors file a record of corner rehabilitation aka, preservation, whenever a PLSS corner is changed. These records are typically filed at county courthouse or county surveyor's office on paper forms, although some counties scan the forms into a PDF or graphic image. Searching the public records can be challenging for a number of reasons. The first issue is that for most jurisdictions, the surveyor must physically go to the public office to search for and make copies of the records. In small jurisdictions the public offices may open only part-time which limits access. Some counties are very large geographically and the courthouse or county surveyor's office may be many miles from the job site. Additionally some public offices put all the records into a cardboard box, some file the records in books sorted by location (township and range), some store the records chronologically by when the record was submitted, and some offices put such a low priority on the corner records that they may not know where to find them.

Figure 36 GCDB point attributes.

A solution for the problem of providing access to public records is to computerize those records and put them online. Creating a GIS layer for corner records simplifies database searches and spatial searches, and makes it possible to create maps that show which PLSS corners have corner records. With today's technology, this is relatively simple. The most logical route is to use GIS in order to provide an additional means to search for records - a means based on location. Location based searches can be performed with web mapping tools and can also be made available using a smart-phone type of application which can tell a surveyor where the nearest corners are based on the surveyor's present location, and can provide a link to the corner record document(s) any time of day, anywhere there is cell coverage. The technology solutions are feasible and developed relatively quickly. The main obstacles to achieving this are the cost and effort required to convert the paper documents to digital form and creating a searchable database for the scanned images, and the political resistance to change. Ideally, a statewide solution is the most logical in order to make the document form and the database consistent across a state. A statewide database also simplifies access for everyone. A statewide database ensures uniformity of corner records and the database, and makes it simple to develop an online form for submittal; an online submittal form saves the surveyor time, ensures uniformity of content and instantly makes the record public to everyone.

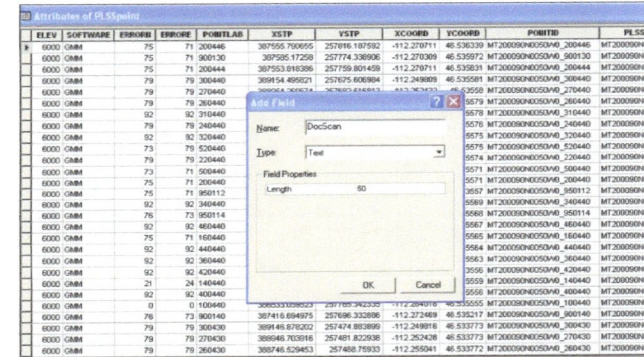

Figure 38 Creating a new field in the GCDB point table.

A GIS layer that represents the locations for corner records is one piece of the puzzle. Here we explain a couple different ways to create a GIS layer for corner records. One method is to create a unique GIS point layer where each point represents a corner record and each point will contain associated attributes and a link to a scanned image. Another method to create a GIS layer is to perform a virtual database *join to* an existing point database using the GCDB. A database join takes advantage of preexisting geometry for the graphic and spatial location, and attaches additional tabular to certain point records based on a unique value in a tabular field.

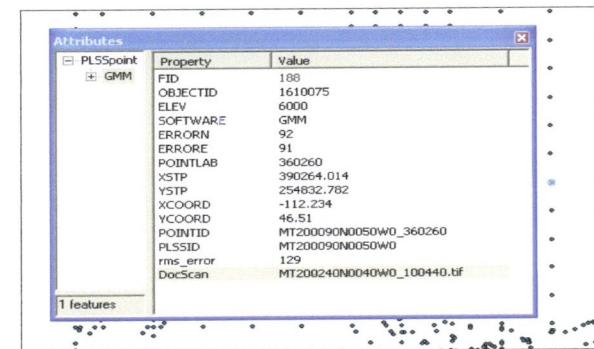

Figure 37 GCDB points attributes – Document name field

Method 1 create a GIS point layer

Overview of steps

1. Scan the documents into digital form - PDF is best because a single PDF file can contain multiple pages

2. Create a GIS point layer with at least one attribute for the scanned document file name.

3. Alternatively, one may add additional fields to facilitate database queries (examples include township, range, section, date, surveyor name, corner name (using the GCDB ID).

All existing corner records that are not already in digital form must be scanned and the digital files must be uniquely named. One file naming strategy is to name each scanned image file using the unique GCDB ID. Since there could be multiple records per PLSS corner, you should use a suffix of some kind to differentiate subsequent records. Adobe PDF (portable document format) is my preference for the digital format because PDF documents allow multiple pages per document. By using PDF you can organize multi-page records, the

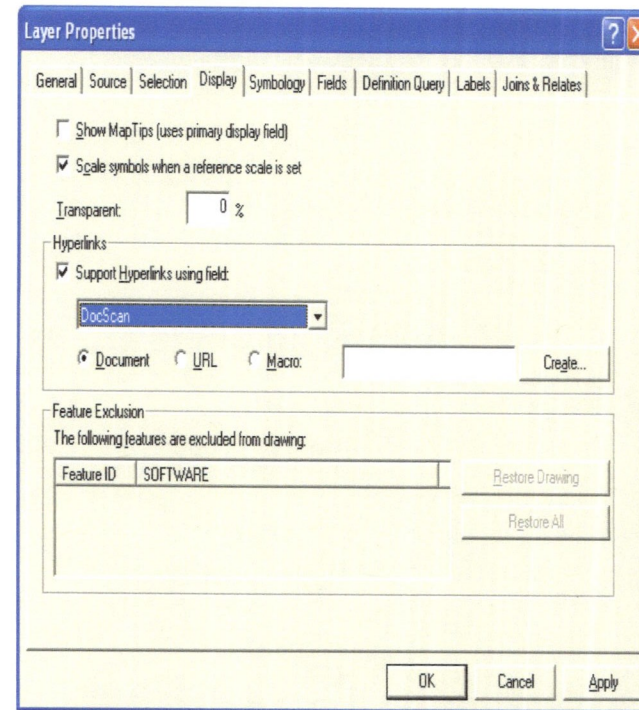

Figure 39 Defining the field to use for hyperlinks to images.

image quality can be high, and the document can be zoom-able. The Adobe Reader also gives the user a lot of control over how to

print a document.

Creating a GIS point for each corner record requires editing in GIS

software - one can simple create a point for a corner record by clicking

on the correct location on a map. Ideally, though, this should be done

by snapping new points to the GCDB point, or by *copying* the

appropriate GCDB GIS point then pasting that point into the corner

records GIS point layer. The latter method of copying GCDB points has

the advantage of bringing along the other GCDB point attributes, some

of which can be helpful when performing queries of the database. This

method is labor intensive - someone has to perform the GIS editing and

ensure that the points are correctly placed on the map.

The next step after each point is created is to type in the scanned file

name as an attribute for the point. The file name attribute will be used

later to create a hyperlink to the corner record document. Note that

every time that a new corner record is created someone must create

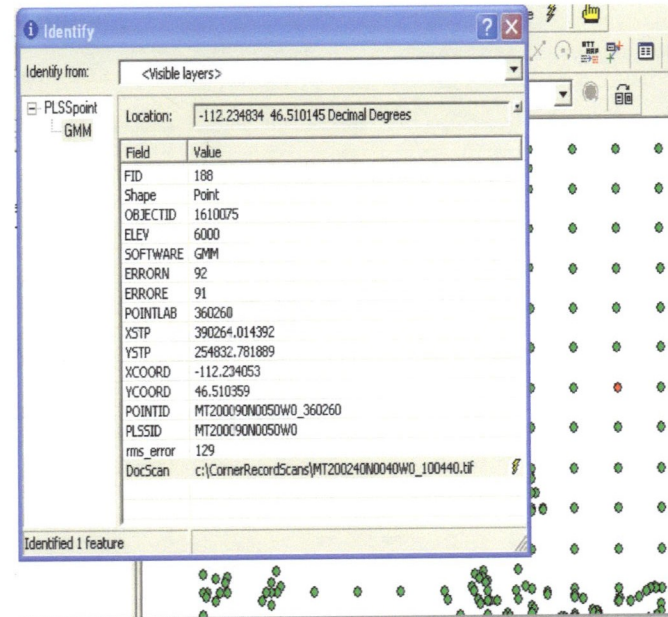

Figure 40 Click on the DocScan filename value to open the scanned image.

the GIS point and add the file name attribute. This effort may be centralized at a responsible agency. Alternatively, the surveyor who

creates the record could do this task automatically when he or she creates a new corner record by using a distributed web-editing

environment.

Method 2 *Create a database table then join the table to the GCDB point layer.*

Overview of steps

1. Scan the corner record documents into digital form as described above.

2. Create a simple database that has one field for the scanned document file name and another field for the GCDB point ID.

3. Join the corner record table to the GCDB point layer using the GCDB point ID as the join field.

The database table must contain a field for the fully unique GCDB ID and the values entered in that field for each corner record, must match the corresponding values in the GCDB point dataset exactly so that the software can perform the table join correctly. In order to recall the scanned corner record document, the database table must also contain a field for the scanned document image file name which can be used as a hyperlink to recall the scanned document file. Depending on how the system is setup it may also be necessary to include the file folder name as well. In addition to the two required fields, the database may also contain other fields as appropriate, such as recordation date, surveyor's name, etc. which may be helpful when performing database queries.

Figure 41 The operating system default application for viewing images of this type will open the document referenced in the hyperlink.

Case Study – Montana Idaho Control Point Database Application

The Idaho-Montana Control Point Database application (http://gisservice.mt.gov/MCPDviewer) is an example of a statewide web application that provides a means to take in information from surveyors. The application connects to the states' surveyor registration list to control who may create an MCPD submittal account. The MCPD has an online control point submittal form that enforces content completeness and continuity. A database administrator, who is a licensed surveyor, then reviews the submittals for form and logical consistency. The administrator then publishes the data to the MCPD viewer. Using the MCPD viewer anyone can search for control points based on querying the database (e.g. by city, or by township, or by who submitted the points,

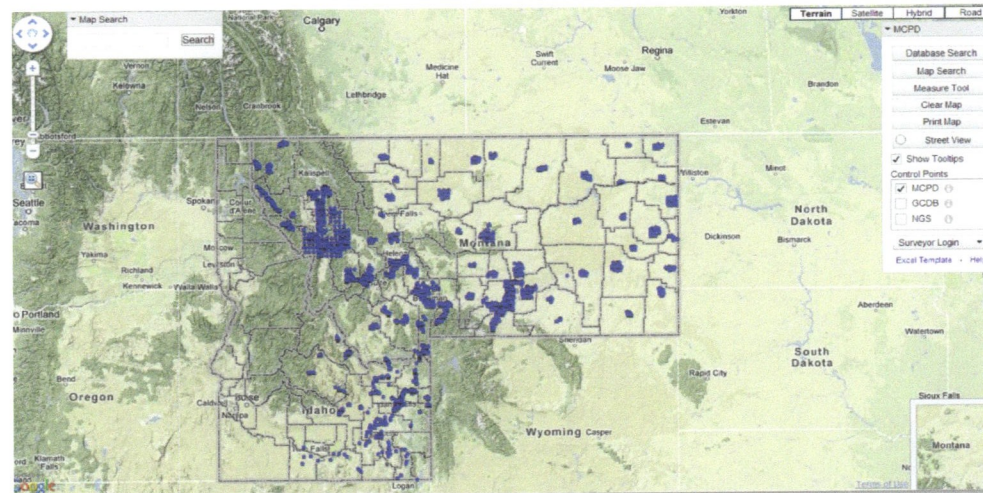

Figure 42 Multistate Control Point Database online viewer.

etc.) or by querying the map by graphically selecting points or performing a search buffer. The MCPD methods for collecting information from surveyors and publishing surveyor information can easily be adapted to other types of surveying related information, such as corner records.

Chapter 4

Ch. 4 How to Improve the GCDB

In this chapter we discuss how to improve the quality of the GCDB spatial alignment through additional surveying, and how to adjust GIS layer when the GCDB coordinates change.

Surveying to Improve the GCDB

 To improve the GCDB, one must develop coordinates that are more accurate - usually by field observations (such as GPS surveys) or by entering, recent, (and presumably more accurate) survey measurements. After entering new coordinates into the GCDB, the GCDB is readjusted to fit the new control, and then any GIS layers tied to it are readjusted to fit the new GCDB. Surveyors and survey measurement data are the key to this process. Surveyors perform the field survey to generate new coordinates, or may even already have coordinates on some Public Lands Survey System corners (tied to the National Spatial Reference System). Surveyors also perform the GCDB re-adjustments, under the guidance of the BLM. The general process is described as follows.

Research and Planning

The first step is to research corner records, certificates of survey & plats, and BLM records to identify PLSS corners that are suitably monumented and that are likely candidates for coordinating with GPS. Common practice is to completely surround a project area with survey control in order to constrain the adjustment within the project area, and to be able to control the adjacent townships. All corner records retrieved are scanned into TIFF format, given the appropriate GCDB ID, and the relevant information added to a geodatabase and referenced to the corner location so that searches are readily performed, and maps easily produced showing the location of monumented corners that have corner records associated with them. One may send out a request for information to local surveyors and government agencies to ask whether they have existing survey control on the PLSS that is tied to the National Spatial Reference Systems (NSRS). If they do and, if they are willing to contribute their control to the project then their coordinate data can also be used to control the GCDB re-adjustment. In some instances, such as when no reliable monumentation exists or access is cost prohibitive, then data entry of existing recorded survey, plat, and deed information may be more a cost-effective means to improve GCDB coordinate values.

As a planning tool, a map is produced showing where possible PLSS corner candidates are, overlain on a map showing the GCDB reliability estimates (need), the location of existing control coordinate data, the suitability for field ties of PLSS corners (existing corner record indicating a suitable monument in locations that will strengthen the GCDB township, and minimal access and safety issues). Based on that information a survey plan is developed

Survey Steps

The next steps are to perform the survey fieldwork (new corner ties, verify other's coordinates, if any).then reduce, calculate and analyze the survey data. Field surveys are performed primarily using Global Positioning Satellite survey methods to obtain one meter or better accuracy on control points. All coordinate information obtained is referenced to the National Spatial Reference System (NSRS) and entered into the GCDB control files as geographic coordinates (NAD 27).

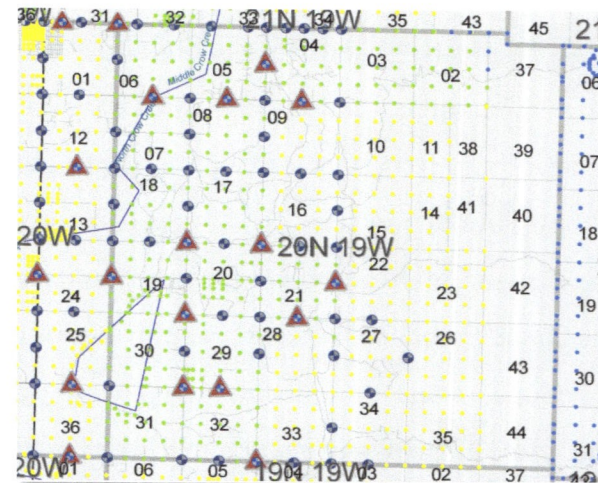

Figure 43 Survey Plan Map – triangles indicate which PLSS corners to GPS. Blue circles indicate which corners have updated record information available. Colored dots indicate the spatial accuracy of GCDB points.

All new and existing control coordinate data are entered into a GCDB control file and each GCDB township is re-adjusted using the BLM's GMM software and approved BLM methods and procedures. The adjustment is analyzed for errors and blunders and re-adjusted if necessary, then submitted to the BLM for review. The BLM then incorporates the re-adjusted GCDB into the agency's framework database.

It is important to note that GCDB error estimates are somewhat subjective, so some estimates may be overly optimistic. Because of this, it is possible for the error to appear to increase after the adjustment, even though the coordinates may be <u>more</u> reliable. In these instances, the accuracy <u>is</u> better but merely appears to have worsened because the original estimate overstated the accuracy. However, it may also be true that the accuracy got worse, which does happen in some areas due to bad survey data or blunders in the GCDB. These errors can usually be isolated and measure taken to improve them. Generally, though one will see an improvement in the GCDB accuracy after new data are entered and the GCDB is adjusted (see charts on the next page). That is, after all, the point.

The charts below summarize changes in GCDB error estimates in townships after the adjustment of every GCDB points' coordinates.

The adjustment, used improved coordinates (accurate to ~ 3 cm) for about a dozen points per township

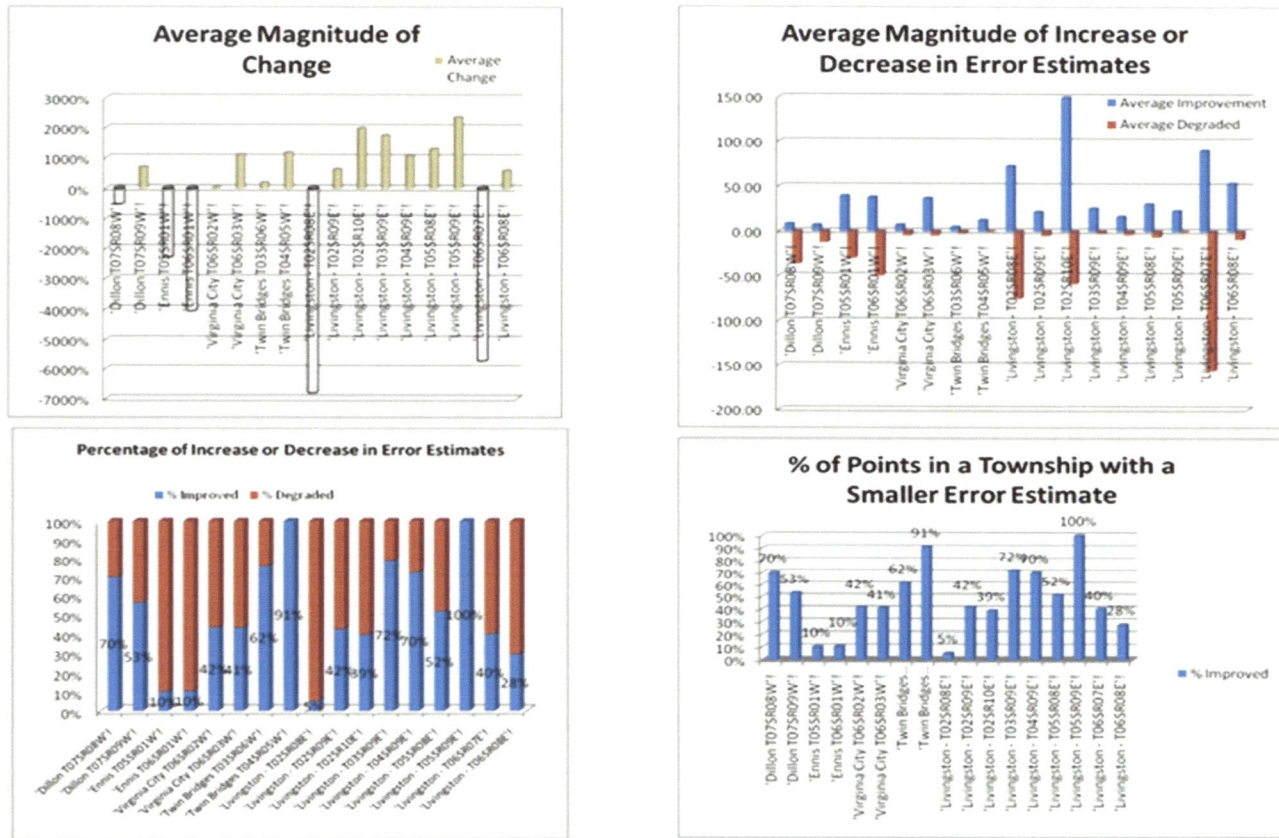

Figure 44 Charting the changes in GCDB error estimates.

Post adjustment analysis – understanding the effect that the new control coordinates have on the townships.

Case Study Example: Montana Experience Improving the Spatial Accuracy of the GCDB

Because Montana based most of its cadaster upon the Bureau of Land Management's Geographic Coordinate Database (GCDB) (a digital representation of the locations of the Public Lands Survey System corners), the process for improving the spatial accuracy of the GIS cadaster first requires upgrades to the accuracy of the GCDB. Montana's surveyors play a significant role in the GCDB enhancement projects where the ultimate goal is to improve the spatial alignment of GIS layers. This article discusses the surveyor's role in making GIS better.

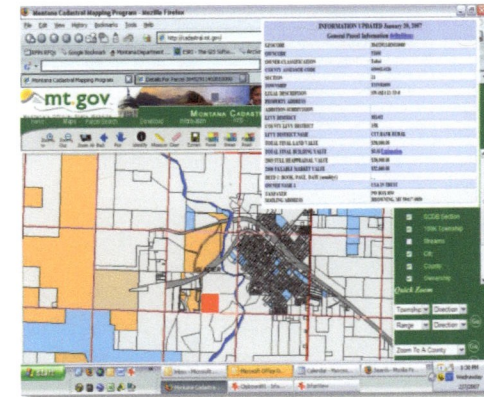

Figure 45 State of Montana Cadastral website provides access to parcels and land ownership. Parcels were built on the GCDB.

Background:

Although Montana is the fourth largest state in land area at more than 145,552 square miles, and has a population density of about 6.2 people per square mile, it is one of the few states in the nation to have a statewide cadastral GIS. The number of parcels in the Montana Cadastral dataset is about the same as the number of people in the state – a little more than 900,000. The annual economic benefits of the Montana Cadastral dataset average in the million dollar range - an obvious demonstration of the value of this system. The benefits accrue from the variety of applications that one can do against a statewide cadastral dataset and the savings in time and money realized by the availability of the data via the World Wide Web (www.cadastral.mt.gov). Nevertheless, the methodologies used to create this statewide cadaster in a few short years in the 1990's resulted in less than optimum spatial accuracy in some areas of the state. "Less than optimum" means that certain applications fail in some areas, such as GPS derived address points not falling in the correct parcel or calculations for impermeable area are difficult to apply uniformly due to the spatial misalignment between GIS layers. Fortunately, spatial accuracy of GIS layers can be improved and surveyors are part of that process. Montana has a spatial accuracy improvement program dedicated to enhancing the spatial accuracy of the cadastral layer in problem areas.

Sources of Errors

Errors arise from a variety of causes. Because Montana based its cadastral framework layer on the Bureau of Land Management's Geographic Coordinate Database (GCDB) in most areas, the accuracy of the parcels is dependent on the accuracy of the GCDB, thus the parcels can be no more accurate than the GCDB. The BLM constructed the GCDB by digitizing coordinates off the US Geological Survey's Topographical maps, supplemented in some areas, by GPS derived coordinates and/or data entry of recent recorded survey bearings and distances. The majority of coordinate pairs of the GCDB are approximations at best. Additionally, other sources of error lie in the reliability of the source data (assessor maps that were hand drafted over the course of decades), and the processes used to digitize the parcels. The digitizing process may have inadvertently introduced errors.

Figure 46 Misalignment of parcels

In practice, best results come from minimizing the error in areas where the density of parcels (and other GIS layers) is highest, that is, the higher the parcels density is, the smaller the spatial error should be. Nevertheless, in Montana, the spatial error varies somewhat randomly so we have some urban areas (high density) with very large errors and some rural areas (such as where one whole section might be a single parcel - more or less one square mile) with relatively high accuracy. Regardless of the source of error, one can improve the spatial accuracy in a couple of ways. For rural areas, which are the majority of the state, the process to improve the parcels starts by improving the GCDB

Re-adjusting GIS Data to an enhanced GCDB

Introduction

When the GCDB coordinates are adjusted through an accuracy enhancement the GIS data that are registered to the GCDB must subsequently be re-adjusted to conform to the new GCDB.

Because the coordinates of the PLSS points in the GCDB have changed, the GIS vertices that had been snapped to those points must now be moved to the new locations. This can be done manually by editing each vertex and snapping it to the new location of the corresponding GCDB point, but for large data sets, it's more efficient to do this programmatically. There is a script that ESRI's Tim Hodson wrote, to perform that function. The GCDB/GIS adjustment script operates on the principle of linking the positions of the original GCDB point ids with new GCDB point ids and moving the associated geography of the feature class or classes to the new GCDB locations. The script calculates the vectors between the old locations and the new locations and uses those vectors to perform a rubber sheeting type of adjustment on the GIS data to the new positions.

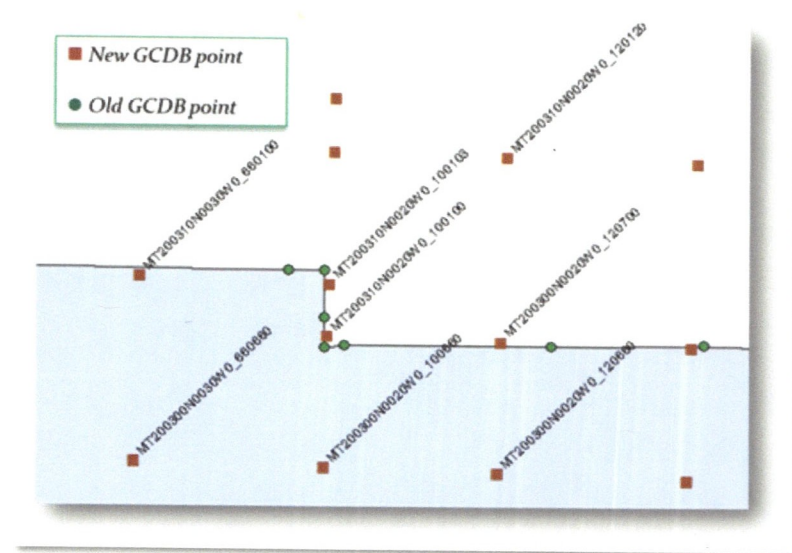

Figure 47 Changes in GCDB coordinates after accuracy upgrade.

The essential elements needed to perform this adjustment, are

- 2 point files – old GCDB points & new GCDB points.

- GIS layer(s) to adjust. These must be in the same spatial reference as the point files.

- The adjustment script. This is Visual Basic script and is available for free download on the GISforSurveyors.com website: http://gisforsurveyors.com/index_htm_files/gcdbAdjustmentScript.txt .

The components of the point files that are needed for this to work are,

✓ The GCDB_ID field which must be present in the old point file and the new point file. This field is the basis for identifying the point in the new point file that corresponds to the point of the same name in the old point file. The field name must be the same in both point files.

✓ Duplicate GCDB_ID *values* in the old point file and the new point file. The script will find the change in coordinates for each point in the old point file that has a corresponding ID in the new point file. These coordinate differences define the adjustment vectors for the rubber sheeting process.

The adjustment will move the GIS vertices that are associated with a GCDB in the old point file to the new location of that corresponding point in the new file. So those vertices will essentially "snap" to the new point locations. Since this is a rubber sheeting procedure, any vertices in the GIS data that are not associated with a GCDB point will also be moved, but that movement will be proportional to its distance between points that

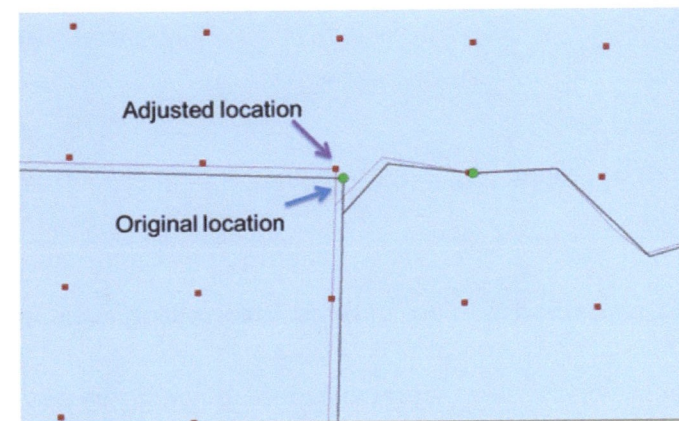

Figure 48 Original vs adjusted positions of GIS feature vertices.

are tied to the GCDB. That is, the movement of the vertices that are not tied to the GCDB will be unpredictable.

Note, that with this adjustment process, it is possible to "fix" certain positions in place because any point in the new point file that has the *same* coordinates as the old point file will have zero displacement, i.e. that point will not move.

In most cases you should inspect the results of the adjustment to make sure that the GIS data moved correctly. You may even need to manually edit some parts after the adjustment.

Transition Area between PLSS and Non-PLSS Boundaries in GCDB Adjustments

Some additional editing might be necessary after the adjustment when non-PLSS features are boundaries because the linkages from the PLSS to the non-PLSS may be modified by the adjustment. Careful inspection of the adjusted data is very important for data integrity. For example, where a boundary follows a section line to a river as shown in figure 49 below, the boundary line coming into the river may move to the new GCDB location. This sometimes creates a new vertex and new line as shown in the detail in figure 49.

Adjusting the PLSS portion to the GCDB, but holding the river boundary fixed, may introduce geometry changes such as shown in figure 49. These types of transition areas must be inspected and dealt with on a case by case basis, and the procedure to deal with the geometry will depend on the descriptions of the boundary.

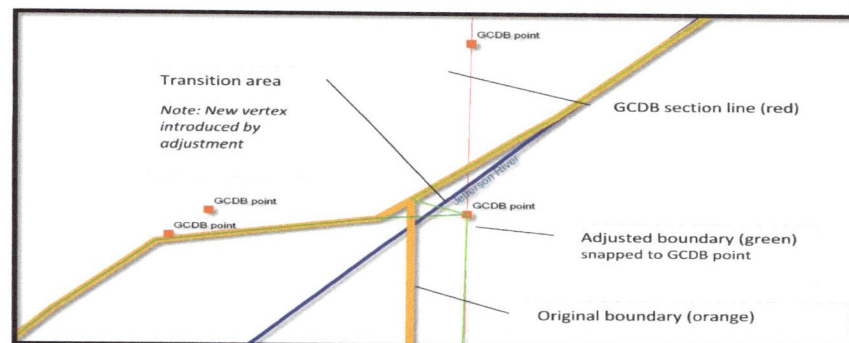

Figure 49 Transition areas between PLSS based and non-PLSS based boundary elements after adjustment.

Application Requirements

These are the operation requirements for using the Cadastral Adjustment utility as of the writing of this document.

1. All datasets – control points, and GIS data to adjust, must have the same spatial reference and units, etc.

2. Editing must be active on the data workspace.

3. There must be two control point files:

    a. Old Control

    b. New Control

4. Control point files must both have a GCDB ID field with matching strings.

Other Notes

1. LARGE FILES:

    Adjustments that use large control point files can sometimes take hours to process.

2. TEST AREAS:

    We recommend performing a test adjustment on a small but representative area of your project data, prior to running a full adjustment against en entire data set. This could identify any issues or anomalies that might arise.

OFFSET CORNERS:

Along certain township lines (standard parallels aka *correction lines)* there may be offset corners in the PLSS plats. In some instances, these offset corners were <u>not</u> digitized in GIS representations of the PLSS, although they are present in the GCDB. The offset corners appear as double corners along the township line; however, the offset corners represent one set of corners for the sections of the township to the south and another set of corners for the sections of the township to the north. Boundaries that follow township lines along the standard parallels, should respect the appropriate offset corners. When one moves a GIS boundary layer from a PLSS data set that only had one corner where there should be two, one should inspect the boundaries along those lines to ensure they are snapped to the appropriate offset corner in the GCDB.

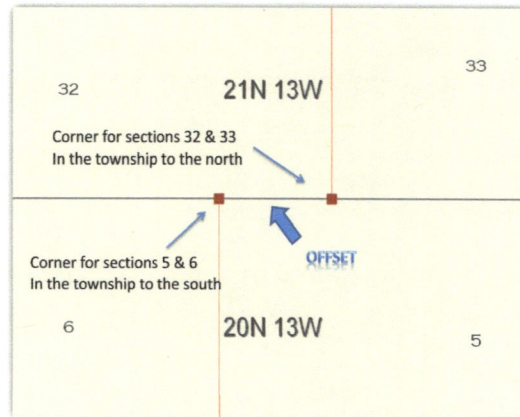

Figure 50 Double corners may occur typically at the north and south township lines. These corners may be offset from each by a few feet to many 10's of feet. The GCDB will have double corners where such corner were surveyed, but PLSS maps at small scales may not

Chapter 5

CH. 5 GCDB Resources

This chapter contains some information on where to obtain GCDB data and additional help and information.

Where to get the GCDB data:

The BLM official website (http://www.blm.gov/wo/st/en/prog/more/gcdb.html) for the GCDB is called the Land Survey Information website. This site features information about the GCDB and includes links to the Standards (http://www.blm.gov/wo/st/en/prog/more/gcdb/blm_gcdb_standards.html) and a map interface for selecting townships to download. My recommendation is that everyone's first stop be the Standards page in order to get a detailed look at what kinds of information are included in the GCDB, and to get descriptions of the origin and appropriate uses for the GCDB.

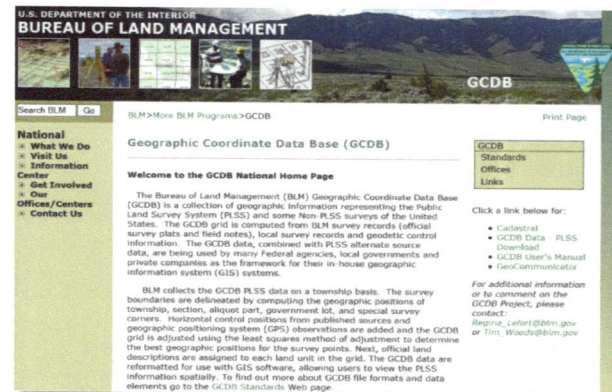

Figure 51 BLM's GCDB website.

Download GCDB (http://www.geocommunicator.gov/blmMap/MapLSIS.jsp)

From the Land Survey Information website one can download GCDB files for townships for any of the western states. The user may select townships by performing a query or by selecting townships with the map interface. The map interface shows which townships are completed, and which are not available.

Once a township (or more) has been selected, the user may choose one of two formats to download: flat files, which are the traditional GCDB format, or the GIS format as shapefiles or a geodatabase. The map interface shows which townships have GIS validated files available. The shapefile format facilitates using

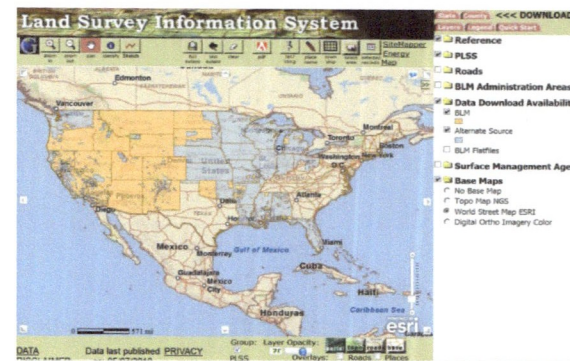

Figure 52 BLM's GCDB download website.

the GCDB in a GIS. Once download and unzipped, the GIS format files can be immediately added to a GIS project, with no further processing required. The GIS format for a township includes a number of files, which represent the PLSS as points, lines and polygons of the various GCDB features. The shape file for the coordinates is a point file which contains attributes for X and Y coordinates. One can use these to obtain estimates of the relative accuracies of GCDB within a township. The graphic below shows the accuracies of each of the GCDB points (which include all points on the PLSS grid – not just township and section corners).

GCDB Training Help

GIS for Surveyors

This is a training and education website about GIS geared toward surveyors. At GISforSurveyors.com you will find information about how Geographic Information Systems technologies relate to the activities of land surveying. Presented there are topics of interest to land surveyors and people involved in GIS. This website provides learning opportunities for land surveyors to better understand GIS, and provides materials for people who are using GIS to learn more about land surveying and survey information as it relates to GIS.

http://GISforSurveyors.com

GCDB Data Model details

Nancy Von Meyer and Fairview Industries publish a detailed document of the Handbook for Standardized PLSS Data (CADNSDI).

http://www.fairview-industries.com/webdocs/Publication%20CADNSDI%20Handbook.pdf

GCDB User's Manual

The Bureau of Land Management publishes a technical manual that documents the structure and domains for the GCDB geometry and attribution.

http://www.blm.gov/co/st/en/BLM_Programs/geographical_sciences/cadastral/gcdb/gcdb_users_manual.html

Glossary

ALIQUOT PARTS – The standard subdivisions of a PLSS section by quantities, e.g. 160 acres or 40 acres, etc. that results from splitting the rectangular geometry in halves, quarters, etc.

ATTRIBUTES – The text and numeric database values associated with GIS feature geometry.

BLM - Bureau of Land Management of the United States of America, Department of the Interior.

CAD- Computer Aided Drafting (or Design) is the computer software technology for creating digital objects.

CADASTRAL – Of or pertaining to cadaster which is a system of records and descriptions of land ownership and interests.

CADASTRAL ADJUSTMENT SCRIPT - As used in this document, the GCDB (or Cadastral) Adjustment script refers to a Visual Basic script written by Tim Hodson of ESRI, that performs a rubber-sheeting adjustment of GIS layers based on two control point files.

CLOSING CORNER - is a corner established where a survey line intersects a previously fixed boundary at a point between corners.

COGO- Coordinate Geometry is a geometric means of creating GIS features, such as drawing lines using coordinates of end points, or by calculating coordinates of endpoints from the measure and angle of a line (e.g. bearing and distance).

CONTROL POINT – A fixed location with coordinates that are known with high reliability.

GCDB - The Geographic Coordinate Database is the Bureau of Land Management's computer representation of the Public Lands Survey System. The GCDB is based on coordinates for each PLSS corner. PLSS corners are angle points, points on line, meandered points, reference points and other points that define the shape and location of the PLSS rectangular system, and may also include mineral surveys, homestead entries, meanders along navigable waterways, and other survey information.

GIS – GEOGRAPHIC INFORMATION SYSTEM- Computer hardware, software, people, processes, and data that comprise a graphical interface to information that has a location component.

GPS – Global Positioning Satellite System is satellite and ground based system of hardware and software used to locate positions on the earth via GPS receivers. Positions are calculated based on the length of time that the GPS signals take to go from the satellites to the GPS receiver.

MEANDERS- The approximate location of water bodies (such as lakes, rivers, swamps, etc.) as mapped during the original Public Lands Surveys. The meanders do not necessarily depict water body edges as they are today, nor do the meanders have any legal bearing.

MISCLOSURE- The failure of survey geometry to form a completely closed polygon due to the starting and ending points having differing coordinates.

OFFSET CORNERS- Duplicate corners of Public Lands Survey System where the ending points of one township are located in a different location than the corners of the adjacent township where in most instances the two corners are coincident. Most offset corners occur along the north or south lines of townships, but may be in other places as well/

NON-GCDB - A non-GCDB PLSS layer is one that is not based on the GCDB. This could be a PLSS representation from the US Census Bureau TIGER files or one that was created by an agency that digitized PLSS tic marks on a USGS Topographical Map Series.

NON-PLSS - In the context of this document, this means any boundary that is not directly referenced to the PLSS. Typically this would be metes and bounds descriptions, roadways, rivers, and other features that are boundaries.

PLSS – Public Lands Survey System is the land allotting system design by President Thomas Jefferson for the orderly disbursement of the western public lands. Also known as the rectangular system, the PLSS is comprised of township blocks approximately six miles by six miles square that are further divided into sections of land that are approximately one square mile in area.

RED LINE AREAS - Geographic areas that rely on non-GCDB mapping controls. Redline areas are typically urban zones of high parcel density and where cadastral geometry have only tenuous connections to the PLSS. In the red line areas other GIS features such as surveyed subdivision corners, road centerlines, sidewalk perimeters, aerial photography, etc. provide the mapping controls for cadastral feature placement and constraint.

REGISTER to the GCDB – As used in this document, a data set that is *registered* to the GCDB is one for which the GCDB is used as the *mapping control,* that is, the GIS layer has at least some of its features coincident to the GCDB points, lines, or polygons. For example, if a parcel boundary follows a section line, and the GIS feature of that parcel uses GCDB section corner points or GCDB section lines, or GCDB section polygons was used to define where that section line is located, then that parcel is registered to the GCDB. This could be done by snapping the parcel vertices to the GCDB points, or by using a GCDB line to create part of the parcel boundary.

TIGER - Topologically Integrated Geographic Encoding and Referencing system are GIS products that were produced by the United States Census Bureau as a mapping and analysis aid for collecting and understanding population information within the United States of America/

Topology- As used in GIS, is the measure of connectivity among features of a dataset or between features of differing datasets.

WGA - Western Governors Association - The Western Governors' Association is an independent, nonprofit organization representing the Governors of 19 states and three US-Flag Pacific islands. Through their Association, the Governors identify and address key policy and governance issues that include natural resources, the environment, human services, economic development, intergovernmental relations and international relations.

Index

Index

ABOUT THE AUTHOR

Rj Zimmer, PLS

M. Eng. Geomatics Engineering

Mr. Zimmer has worked with the GCDB since the late 1990's for parcel mapping. He has used the GCDB to geo-register GIS data, and he has been involved with adjusting the GCDB and multiple PLSS based GIS datasets. Mr. Zimmer also developed strategies for automating GIS data adjustments to GCDB data.

Rj Zimmer is a land surveyor, GIS consultant, educator, and writer. He has published more than sixty articles on GIS for surveyors. Mr. Zimmer is a registered **professional land surveyor and geomatics consultant located** in Helena, Montana. He is a two-time recipient of ESRI's Award for Special Achievement in GIS and received Montana's Distinguished Service Award from the Montana Association of GIS Professionals.

Mr. Zimmer began surveying in 1977 in Oregon, and has performed construction, forest boundaries, hydrographic, cadastral, topographic, engineering, and right of way surveys throughout the Pacific Northwest. Mr. Zimmer is licensed to practice land surveying in Oregon and Montana.

In the late 1980's Mr. Zimmer began working with land survey information systems and geographic information systems, and has since developed hundreds of databases and systems for private and public agencies.

Mr. Zimmer is a writer for **The American Surveyor** Magazine, (http://www.amerisurv.com/content/category/17/191/136/), and has published in the Surveying and Land Information Systems (SALIS) journal, and Professional Surveyor Magazine. He has designed training materials and user's manuals, presented workshops and seminars on topics related to land surveying and Geographic Information Systems, and the Global Positioning System.

Mr. Zimmer teaches Geomatics courses as adjunct faculty to **Carroll College (http://carroll.edu/),** in Helena Montana. He also chaired the Montana Geodetic Control Working Group, was a co-founder of the Montana GPS User's Group, chaired the Montana Local Government GIS Coalition, and was a member of the Montana Governor's Council on GIS.

You may contact Rj Zimmer at RjZimmer@gisforsurveyor.com.

www.ingramcontent.com/pod-product-compliance
Lightning Source LLC
Chambersburg PA
CBHW041702200326
41518CB00002B/163